# The Concise Book of Muscles

## Third Edition

## Chris Jarmey
## John Sharkey

lotus
publishing
Chichester, England

**North Atlantic Books**
Berkeley, California

First published in 2003, the second edition published in 2008, and this third edition published in 2015 by
**Lotus Publishing**
Apple Tree Cottage, Inlands Road, Nutbourne, Chichester, PO18 8RJ and
**North Atlantic Books**
Berkeley, California

**Anatomical Drawings** Amanda Williams
**Exercise Drawings** Matt Lambert
**Text Design** Wendy Craig
**Cover Design** Jasmine Hromjak
**Printed and Bound** in the UK by Bell and Bain Limited

*The Concise Book of Muscles, Third Edition* is sponsored and published by the Society for the Study of Native Arts and Sciences (dba North Atlantic Books), an educational nonprofit based in Berkeley, California, that collaborates with partners to develop cross-cultural perspectives, nurture holistic views of art, science, the humanities, and healing, and seed personal and global transformation by publishing work on the relationship of body, spirit, and nature.

North Atlantic Books' publications are available through most bookstores. For further information, visit our website at www.northatlanticbooks.com or call 800-733-3000.

**British Library Cataloguing-in-Publication Data**
A CIP record for this book is available from the British Library
ISBN 978 1 905367 62 7 (Lotus Publishing)
ISBN 978 1 62317 020 2 (North Atlantic Books)

**The Library of Congress has cataloged the first edition as follows:**
Jarmey, Chris.
The concise book of muscles/Chris Jarmey.
p. ; cm. ill.
Includes bibliographical references and index.
ISBN 1-55643-466-9 (pbk. : alk. paper)
1. Muscles--Handbooks, manuals, etc.
[DNLM: 1. Muscles--innervation--Atlases. 2.
Muscles--innervation--Handbooks. 3. Anatomy, Regional--Atlases. 4.
Anatomy, Regional--Handbooks. 5. Muscles--anatomy & histology--Atlases.
6. Muscles--anatomy & histology--Handbooks. WE 17 J37c 2003] I.
Title.
QP321.J43 2003
612.7'4--dc21

2002155580

# Contents

# Preface

I was very honored to be tasked with preparing the third edition of *The Concise Book of Muscles*, building on the excellent work that Chris Jarmey produced in the previous two editions. Much has changed since the second edition, but I have endeavored to keep to the concise and easy-reference format that has made this text such a popular resource.

Of course, time moves on and little stays the same—this is true of all things, including anatomy. Time gives birth to new facts, models, and hypotheses worthy of consideration and finally acceptance. New research has emerged on the topic of fascia and living motion, with the expression of new theories on myofascial force transmission and the continuity of our living architecture. While not throwing the baby out with the bathwater, my intention is that this latest edition should be seen as a product of a new understanding of the role of muscles and fascia (or, more correctly, connective tissue) in both force transmission and human movement (or living movement).

To fully understand and appreciate the new and persuasive model of biotensegrity based on muscle synergies and four-bar closed kinematics, we must first be familiar with the fallacy of the old model of origin and insertion—two-bar pin joints and external forces—that underpins modern biomechanics. In order to understand the present and the future, we must understand the past. Today, the study of anatomy is based on a tradition that is hundreds of years old; anatomy was a reflection of the vision and beliefs of those early anatomists. Many names given to muscles bore little or no relation to their functions, but reflected more what the anatomist saw: muscles were therefore named maximus or minimus, longus or brevis, anterior or posterior, etc. Even the word "muscle" originates from Latin *musculus*, meaning "little mouse."

As a clinical anatomist I cherish the history of anatomy and particularly the history which led to the anatomical naming of tissues, organs, muscles, and systems. We now appreciate that no one muscle is responsible for one specific movement, and that the brain does not think in terms of muscles but rather in terms of successfully completing a movement. Let us embrace and cherish the rich history, language, and definitions of anatomy, while appreciating the need for new explanations, new models, and a new understanding of anatomy based on sound scientific principles and continuity.

*John Sharkey MSc, Clinical Anatomist (British Association of Clinical Anatomists)*

# Introduction

## About this Book

This book is designed in quick-reference format to offer useful information about the main skeletal muscles that are central to sport, dance, exercise science, and bodywork therapy. Each muscle section is color-coded for ease of reference. Enough detail is included regarding each muscle's origin, insertion, action, and nerve innervation (including the nerve's common course or path) to meet the requirements of the student and practitioner of bodywork, movement therapies, and movement arts. The book aims to present that information accurately and in a particularly clear and user-friendly format, especially as anatomy can seem heavily laden with technical terminology. Technical terms are therefore explained in parenthesis throughout the text.

The information about each muscle is presented in a uniform style throughout. An example is given below, with the meanings of headings explained in bold (some muscles will have abbreviated versions of this).

## Layers

Throughout this text, the term *layer* is used to describe the anatomy of fascial (connective) tissue or the positioning of one structure relative to another. The use of this term is for convenience, and the sense should not be taken literally—there are no physical layers in the human body. Layers are created when a dissection is performed and tissues are separated by scalpel or blunt dissection. Continuity is key, and everything is connected to everything else.

The name of the muscle.

The attachment that remains relatively fixed during muscular contraction—i.e. the end of the muscle which is fixed to the bone that does not move, thereby acting as an anchor for the muscle to pull its opposite end (insertion) toward this fixed attachment (see p. 23).

The attachment that moves—i.e. the opposite end of the muscle to the origin. Note that when the insertion remains relatively fixed and the origin moves, the muscle is said to be performing a reversed action (i.e. origin to insertion). This occurs often. Generally, the origin is more proximal (toward the center of the body) and the insertion is more distal (toward the periphery of the body).

The movement or effect caused when the muscle contracts.

The nerve that activates the muscle.

Everyday activity/ activities to which the muscle contributes.

A few key examples, although each muscle will be involved in varying degrees in most sports.

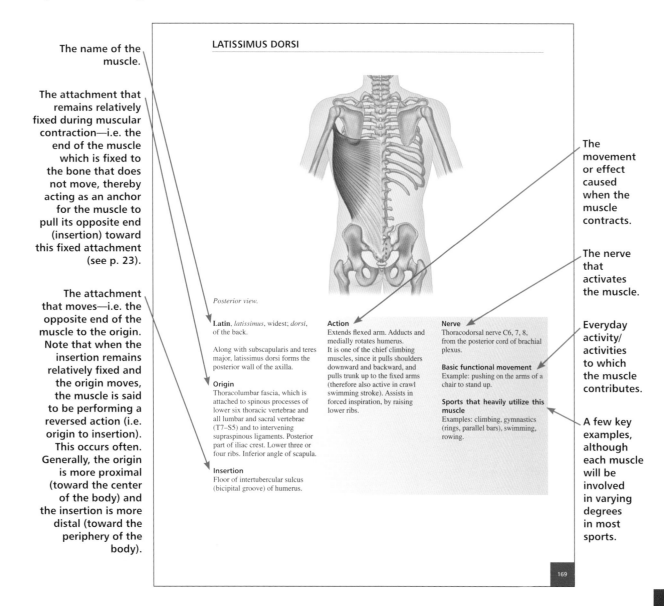

**LATISSIMUS DORSI**

*Posterior view.*

**Latin**, *latissimus*, widest; *dorsi*, of the back.

Along with subscapularis and teres major, latissimus dorsi forms the posterior wall of the axilla.

**Origin**
Thoracolumbar fascia, which is attached to spinous processes of lower six thoracic vertebrae and all lumbar and sacral vertebrae (T7–S5) and to intervening supraspinous ligaments. Posterior part of iliac crest. Lower three or four ribs. Inferior angle of scapula.

**Insertion**
Floor of intertubercular sulcus (bicipital groove) of humerus.

**Action**
Extends flexed arm. Adducts and medially rotates humerus. It is one of the chief climbing muscles, since it pulls shoulders downward and backward, and pulls trunk up to the fixed arms (therefore also active in crawl swimming stroke). Assists in forced inspiration, by raising lower ribs.

**Nerve**
Thoracodorsal nerve C6, 7, 8, from the posterior cord of brachial plexus.

**Basic functional movement**
Example: pushing on the arms of a chair to stand up.

**Sports that heavily utilize this muscle**
Examples: climbing, gymnastics (rings, parallel bars), swimming, rowing.

169

## Peripheral Nerve Supply

The nervous system comprises:

- The central nervous system (CNS)—i.e. the brain and spinal cord.
- The peripheral nervous system (PNS), including the autonomic nervous system—i.e. all neural structures outside the brain and spinal cord.

The PNS consists of 12* pairs of cranial nerves and 31 pairs of spinal nerves (with their subsequent branches). The spinal nerves are numbered according to the level of the spinal cord from which they arise (the level is known as the *spinal segment*). Muscle innervation pathways are discussed in Appendix 1.

In this book the relevant peripheral nerve supply is listed with each muscle, for those who need to know. However, information about the spinal segment** from which the nerve fibers emanate often differs between the various sources. This is because it is extremely challenging for clinical anatomists to trace the route of an individual nerve fiber through the intertwining maze of other nerve fibers as it passes through its plexus (plexus = a network of nerves: from Latin *plectere* = "to braid"). Therefore, such information has been derived mainly from empirical clinical observation, rather than through dissection of the body.

In order to give the most accurate information possible, the method devised by Florence Peterson Kendall and Elizabeth Kendall McCreary has been duplicated in this book. Kendall and McCreary (1983) integrated information from six well-known anatomy reference texts: those written by Cunningham, deJong, Bumke and Foerster, Gray, Haymaker and Woodhall, and Spalteholz. Adopting the same procedure, and then cross-matching the results with those of Kendall and McCreary, the following system of emphasizing the most important nerve roots for each muscle has been used in this book.

Let us take the supinator muscle as our example; this is supplied by the posterior interosseous nerve, a continuation of the deep branch of the radial nerve C5, **6**, (7). The relevant spinal segment is indicated by the letter "C" and the numbers "5, **6**, (7)." Bold numbers, e.g. **6**, indicate that most (at least five) of the sources agree. Numbers that are not bold, e.g. 5, denote agreement by three or four sources. Numbers not in bold and in parenthesis, e.g. (7), reflect agreement by two sources only, or that more than two sources specifically regarded it as a very minimal supply. If a spinal segment was mentioned by only one source, it was disregarded. Hence, bold type indicates the major innervation, non-bold type indicates the minor innervation, and numbers in parenthesis suggest possible or infrequent innervation.

*There are technically 13 pairs of cranial nerves (Fuller, Burger 1990). (The first cranial nerve is *nervus terminalis (NT)*, also known as the terminal nerve, nerve of pinkus, or cranial nerve 0; however, as there is no roman symbol for zero, an "N" for the Latin word *nulla* is the preferred numerical designation.) This means we have 0 to 12 cranial nerves, which gives us 13 in total (counting 0 as the first number). This is important for surgeons, medical professionals, osteopaths, chiropractors, physiotherapists, and manual therapists of every stripe. It has been suggested that cranial nerve XIV (the *nerve of Wrisberg* or the *nervus intermedius* or *intermediary nerve*) is the 14th cranial nerve and not merely an offshoot of cranial nerve VII.

** A spinal segment is the part of the spinal cord that gives rise to each pair of spinal nerves (a pair consists of one nerve for the left side and one for the right side of the body). Each spinal nerve contains motor and sensory fibers. Soon after the spinal nerve exits through the foramen (the opening between adjacent vertebrae), it divides into a dorsal primary ramus (directed posteriorly) and a ventral primary ramus (directed laterally or anteriorly). Fibers from the dorsal rami innervate the skin and extensor muscles of the neck and trunk. The ventral rami supply the limbs, plus the sides and front of the trunk.

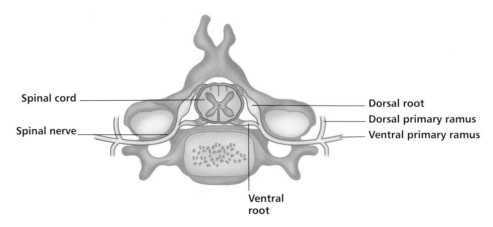

Spinal cord — Dorsal root — Dorsal primary ramus — Ventral primary ramus

Spinal nerve — Ventral root

*A spinal segment, showing the nerve roots combining to form a spinal nerve, which then divides into ventral and dorsal rami.*

# 1 Anatomical Orientation

## Anatomical Directions

To describe the relative positions of body parts and their movements, it is essential to have a universally accepted initial reference position. The standard body position, known as the *anatomical position*, serves as this reference. The anatomical position is simply the upright standing position, feet flat on the floor with the arms hanging by the sides and the palms facing forward (see Figure 1.1). The directional terminology used refers to the body as if it were in the anatomical position, regardless of its actual position. Note also that the terms *left* and *right* refer to the sides of the object or person being viewed, and not those of the reader.

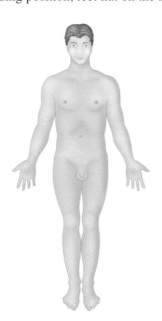

*Figure 1.1. Anterior*
In front of; toward or at the front of the body.

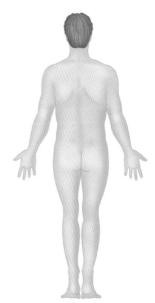

*Figure 1.2. Posterior*
Behind; toward or at the back of the body.

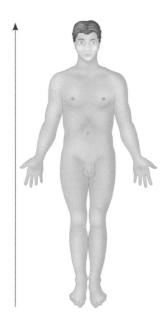

*Figure 1.3. Superior*
Above; toward the head or the upper part of the structure or the body.

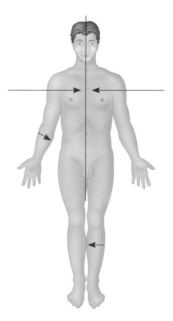

*Figure 1.5. Medial*
(from Latin *medius* = "middle")
Toward or at the midline of the body; on the inner side of a limb.

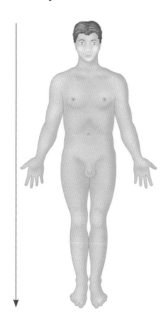

*Figure 1.4. Inferior*
Below; away from the head or toward the lower part of the structure or the body.

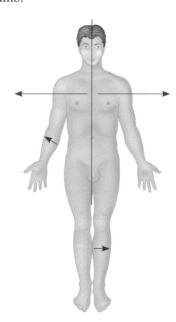

*Figure 1.6. Lateral*
(from Latin *latus* = "side").
Away from the midline of the body; on the outer side of the body or a limb.

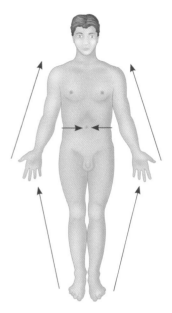

*Figure 1.7. Proximal*
(from Latin *proximus* = "nearest")
Closer to the center of the body (the navel), or to the
point of attachment of a limb to the body torso.

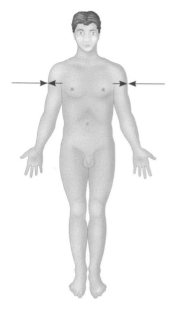

*Figure 1.9. Superficial*
Toward or at the body surface.

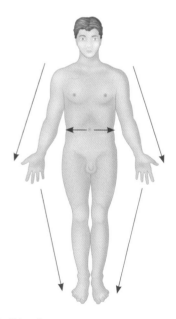

*Figure 1.8. Distal*
(from Latin *distans* = "distant").
Farther away from the center of the body, or from the
point of attachment of a limb to the torso.

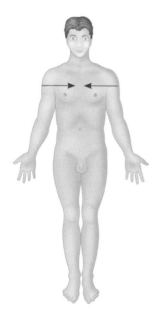

*Figure 1.10. Deep*
Farther away from the body surface; more internal.

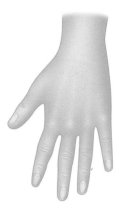

*Figure 1.11. Dorsal*
(from Latin *dorsum* = "back").
On the posterior surface of something, e.g. the back
of the hand or the top of the foot.

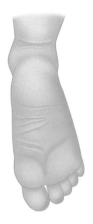

*Figure 1.13. Plantar*
(from Latin *planta* = "sole")
On the sole of the foot.

*Figure 1.12. Palmar*
(from Latin *palma* = "palm").
On the anterior surface of the hand, i.e. the palm.

## Regional Areas

The two primary divisions of the body are its *axial* part, consisting of the head, neck, and trunk, and its *appendicular* parts, consisting of the limbs, which are attached to the axis of the body. Figure 1.14 shows the terms used to indicate specific body areas. Terms in parenthesis are the lay terms for the area.

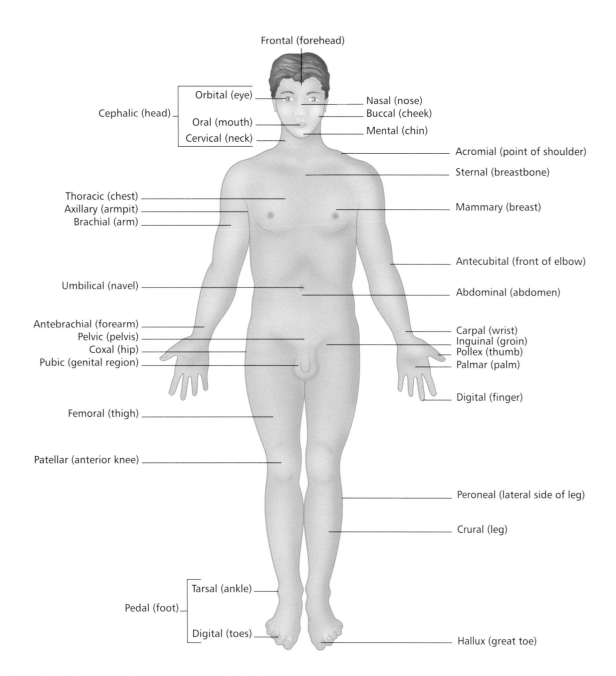

*Figure 1.14. Terms used to indicate specific body areas: (a) anterior view.*

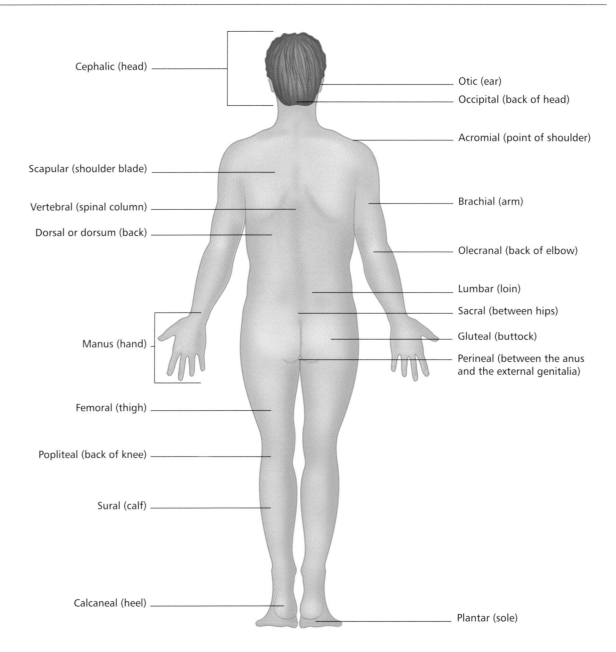

Cephalic (head)

Otic (ear)

Occipital (back of head)

Acromial (point of shoulder)

Scapular (shoulder blade)

Vertebral (spinal column)

Brachial (arm)

Dorsal or dorsum (back)

Olecranal (back of elbow)

Lumbar (loin)

Sacral (between hips)

Gluteal (buttock)

Manus (hand)

Perineal (between the anus and the external genitalia)

Femoral (thigh)

Popliteal (back of knee)

Sural (calf)

Calcaneal (heel)

Plantar (sole)

*Figure 1.14. Terms used to indicate specific body areas: (b) posterior view.*

## Planes of the Body

The term *plane* refers to a two-dimensional section through the body; it provides a view of the body or body part, as though it has been cut through by an imaginary line.

- The sagittal planes cut vertically through the body from anterior to posterior, dividing it into right and left halves. Figure 1.15 shows the midsagittal plane.

- The frontal (coronal) planes pass vertically through the body, dividing it into anterior and posterior sections, and lie at right angles to the sagittal plane.

- The transverse planes are horizontal cross sections, dividing the body into upper (superior) and lower (inferior) sections, and lie at right angles to the other two planes.

Figure 1.15 illustrates the most frequently used planes.

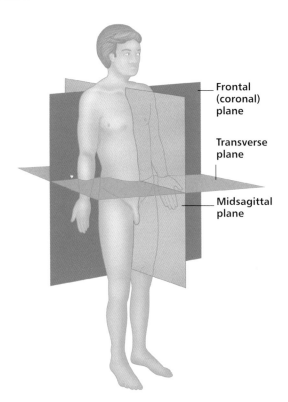

Frontal (coronal) plane

Transverse plane

Midsagittal plane

*Figure 1.15. Planes of the body.*

## Anatomical Movements

The direction in which body parts move is described in relation to the fetal position. Moving into the fetal position results from flexion of all the limbs; straightening out of the fetal position results from extension of all the limbs.

*a)*

*b)*

*Figure 1.16. (a) Flexion into the fetal position. (b) Extension out of the fetal position.*

### Main Movements

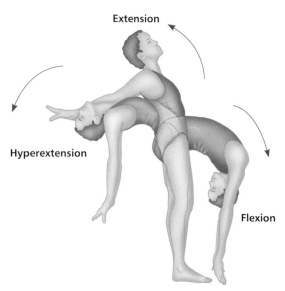

Extension

Hyperextension

Flexion

*Figure 1.17.* **Flexion**: *bending to decrease the angle between bones at a joint. From the anatomical position, flexion is usually forward, except at the knee joint where it is backward. The way to remember this is that flexion is always toward the fetal position.* **Extension**: *to straighten or bend backward away from the fetal position.* **Hyperextension**: *to extend the limb beyond its normal range.*

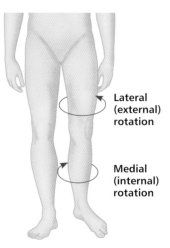

Figure 1.18. **Lateral flexion**: to bend the torso or head laterally (sideways) in the frontal (coronal) plane.

Figure 1.20. **Rotation**: movement of a bone or the trunk around its own longitudinal axis.
**Medial rotation**: to turn inward, toward the midline.
**Lateral rotation**: to turn outward, away from the midline.

## Other Movements

Movements described in this section are those that occur only at specific joints or parts of the body, usually involving more than one joint.

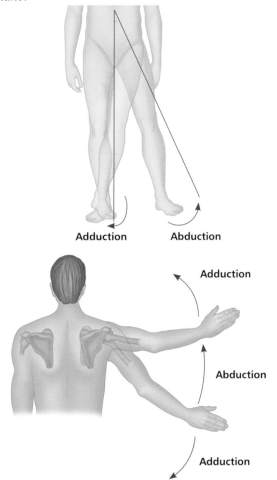

Figure 1.21. (a) **Pronation**: to turn the palm of the hand down to face the floor (if standing with elbow bent 90°, or if lying flat on the floor) or away from the anatomical and fetal positions.

Figure 1.21. (b) **Supination**: to turn the palm of the hand up to face the ceiling (if standing with elbow bent 90°, or if lying flat on the floor) or toward the anatomical and fetal positions.

Figure 1.19. **Abduction**: movement of a bone away from the midline of the body or the midline of a limb.
**Adduction**: movement of a bone toward the midline of the body or the midline of a limb.

Note: For abduction of the arm to continue above the height of the shoulder (elevation through abduction), the scapula must rotate on its axis to turn the glenoid cavity upward (see Figure 1.27(b)).

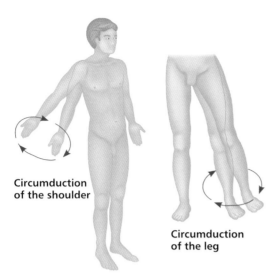

Circumduction
of the shoulder

Circumduction
of the leg

*Figure 1.22.* **Circumduction**: *movement in which the distal end of a bone moves in a circle, while the proximal end remains stable; the movement combines flexion, abduction, extension, and adduction.*

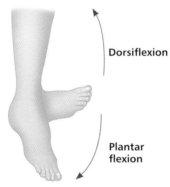

Dorsiflexion

Plantar
flexion

*Figure 1.23.* **Plantar flexion**: *to point the toes down toward the ground.*
**Dorsiflexion**: *to point the toes up toward the sky.*

Eversion    Inversion

*Figure 1.24.* **Inversion**: *to turn the sole of the foot inward, so that the soles would face toward each other.*
**Eversion**: *to turn the sole of the foot outward, so that the soles would face away from each other.*

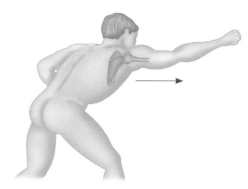

*Figure 1.25.* **Protraction**: *movement forward in the transverse plane—for example, protraction of the shoulder girdle, as in rounding the shoulder.*

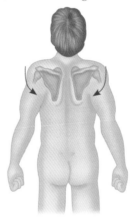

*Figure 1.26.* **Retraction**: *movement backward in the transverse plane, as in bracing the shoulder girdle back, military style.*

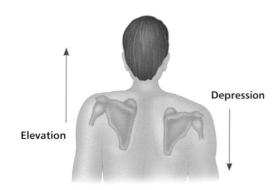

*Figure 1.27. (a)* **Elevation***: movement of a part of the body upward along the frontal plane—for example, elevating the scapula by shrugging the shoulders.* **Depression***: movement of an elevated part of the body downward to its original position.*

*Figure 1.27. (c)* **Elevation through flexion***: flexing the arm at the shoulder joint, then continuing to raise it above the head in the sagittal plane.*

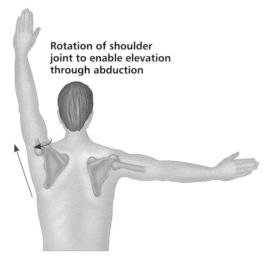

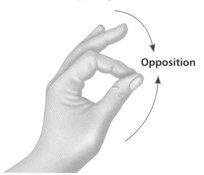

*Figure 1.28.* **Opposition***: a movement specific to the saddle joint of the thumb; it enables you to touch your thumb to the tips of the fingers of the same hand.*

*Figure 1.27. (b)* **Elevation through abduction***: abducting the arm at the shoulder joint, then continuing to raise it above the head in the frontal plane.*

# 2

# Skeletal Muscle, Musculoskeletal Mechanics, Fascia, and Biotensegrity

## Skeletal Muscle Structure and Function

Skeletal (somatic or voluntary) muscles make up approximately 40% of the total human body weight. Their primary function is to produce movement through the ability to contract and inhibit in a coordinated manner. They are attached to bone by tendons (or sometimes directly). The place where a muscle attaches to a relatively stationary point on a bone, either directly or via a tendon, is called the *origin*. When the muscle contracts, it transmits tension to the bones across one or more joints, and movement occurs. The end of the muscle which attaches to the bone that moves is called the *insertion*.

## Overview of Skeletal Muscle Structure

The functional unit of skeletal muscle is known as a *muscle fiber*, which is an elongated cylindrical cell with multiple nuclei, ranging from 10 to 100 microns in width, and a few millimeters to 30+ centimeters in length. The cytoplasm of the fiber is called the *sarcoplasm*, which is encapsulated inside a cell membrane called the *sarcolemma*. A delicate membrane known as the *endomysium* surrounds each individual fiber.

Muscle fibers are grouped together in bundles, or *fasciculi* (fascicles), covered by the *perimysium*. The bundles of muscle fibers are themselves grouped together, and the whole muscle is encased in a fascial sheath called the *epimysium*. These muscle membranes lie throughout the entire length of the muscle, from the tendon of the origin to the tendon of the insertion. The whole structure is sometimes referred to as the *musculotendinous unit*.

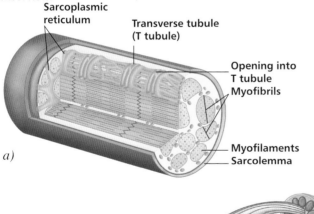

*a)*

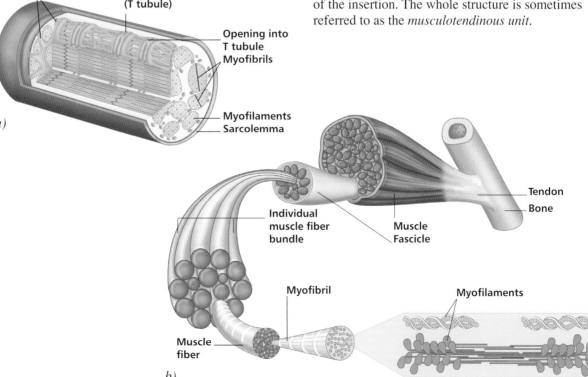

*b)*

*Figure 2.1. a) each skeletal muscle fiber is a single cylindrical muscle cell, b) cross section of muscle tissue.*

In defining the structure of muscle tissue in more detail, from the minute to gross, we therefore have the following components: myofibrils, endomysium, fasciculi, perimysium, epimysium, deep fascia, and superficial fascia.

## Myofibrils

Through an electron microscope, one can distinguish the contractile elements of a muscle fiber, known as *myofibrils*, running the entire length of the fiber. Each myofibril reveals alternate light and dark banding, producing the characteristic cross-striation of the muscle fiber; these bands are called *myofilaments*. The light bands are referred to as *isotropic (I) bands* and consist of thin myofilaments made of the protein actin. The dark ones are called *anisotropic (A) bands*, consisting of thicker myofilaments made of the protein myosin. A third connecting filament is made of the sticky protein titin, which is the third-most-abundant protein in human tissue.

The myosin filaments have paddle-like extensions that emanate from the filaments, rather like the oars of a boat. These extensions latch onto the actin filaments, forming what are described as "cross-bridges" between the two types of filament. The cross-bridges, using the energy of ATP, pull the actin strands closer together.[*] Thus, the light and dark sets of filaments increasingly overlap, like an interlocking of the fingers, resulting in muscle contraction. One set of actin–myosin filaments is called a *sarcomere*.

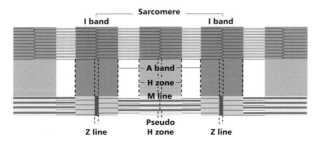

*Figure 2.2. The myofilaments in a sarcomere. A sarcomere is bounded at both ends by the Z line.*

- The lighter zone is known as the *I band*, and the darker zone the *A band*.
- The *Z line* is a thin dark line at the midpoint of the I band.
- A *sarcomere* is defined as the section of myofibril between one Z line and the next.
- The center of the A band contains the *H zone*.
- The *M line* bisects the H zone, and delineates the center of the sarcomere.

If an outside force causes a muscle to stretch beyond its resting level of tonus (see "Tonus" below), the interlinking effect of the actin and myosin filaments that occurs during contraction is reversed. Initially, the actin and myosin filaments accommodate the stretch, but as the stretch continues, the titin filaments increasingly "pay out" to absorb the displacement. Thus, it is the titin filament that determines the muscle fiber's extensibility and resistance to stretch. Research indicates that a muscle fiber (sarcomere), if properly prepared, can be elongated up to 150% of its normal length at rest.

## Endomysium

A delicate connective tissue called *endomysium* lies outside the sarcolemma of each muscle fiber, separating each fiber from its neighbors, but also connecting them together.

## Fasciculi

Muscle fibers are arranged in parallel bundles called *fasciculi*.

## Perimysium

Each fasciculus is bound by a denser collagenic sheath called the *perimysium*.

## Epimysium

The entire muscle, which is therefore an assembly of fasciculi, is wrapped in a fibrous sheath called the *epimysium*; this arrangement facilitates force transmission.

## Deep Fascia (Profundus)

A coarser sheet of fibrous connective tissue lies outside the epimysium, binding individual muscles into functional groups. This deep fascia extends to wrap around other adjacent structures.

## Superficial Fascia (Superficialis)

While its anatomy and topography differs from region to region, providing specialization, the superficial fascia is primarily a fatty fascia that contains oblique septa and connects the skin to the deep fascia. Contractile fibers have been reported in the superficial fascia, particularly in the neck.

---

*The generally accepted hypothesis to explain muscle function is partly described by Hanson and Huxley's sliding filament theory (Huxley and Hanson 1954). Muscle fibers receive a nerve impulse that causes the release of calcium ions stored in the muscle. In the presence of the muscle's fuel, known as *adenosine triphosphate (ATP)*, the calcium ions bind with the actin and myosin filaments to form an electrostatic (magnetic) bond. This bond causes the fibers to shorten, resulting in their contraction or increase in tonus. When the nerve impulse ceases, the muscle fibers relax. Because of their elastic elements, the filaments recoil to their non-contracted lengths, i.e. their resting level of tonus.

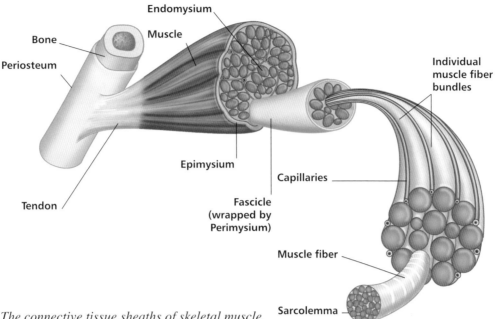

*Figure 2.3. The connective tissue sheaths of skeletal muscle.*

## Muscle Attachment

The way a muscle attaches to bone or other tissues is through either a direct attachment or an indirect attachment. A *direct attachment* (called a *fleshy attachment*) is where the perimysium and epimysium of the muscle unite and fuse with the periosteum of a bone, the perichondrium of a cartilage, a joint capsule, or the connective tissue underlying the skin (some muscles of facial expression being good examples of the last). An *indirect attachment* is where the connective tissue components of a muscle fuse together into bundles of collagen fibers to form an intervening tendon. Indirect attachments are much more common. The different types of tendinous attachment are: tendons and aponeuroses, intermuscular septa, and sesamoid bones.

### Tendons and Aponeuroses

When muscle fascia (connective tissue component of a muscle) combine together and extend beyond the end of the muscle as round cords or flat bands, the tendinous attachment is called a *tendon*; if they extend as a thin, flat, and broad sheet-like material, the attachment is called an *aponeurosis*. The tendon or aponeurosis secures the muscle to the bone or cartilage, to the fascia of other muscles, or to a seam of fibrous tissue called a *raphe*. Flat patches of tendon may form on the body of a muscle where it is exposed to friction. This may occur, for example, on the deep surface of the trapezius where it rubs against the spine of the scapula.

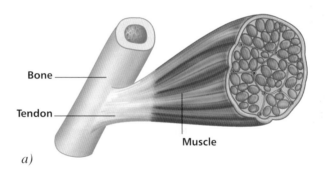

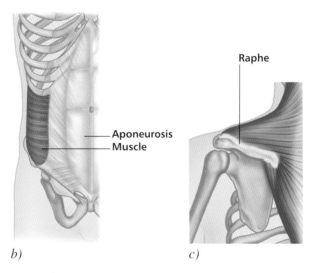

*Figure 2.4. a) Tendon attachment. b) Attachment by aponeurosis. c) Flat patches of tendon on the deep surface of the trapezius.*

## Intermuscular Septa

In some cases, flat sheets of dense connective tissue known as *intermuscular septa* penetrate between muscles, providing another medium to which muscle fibers may attach.

## Sesamoid Bones

If a tendon is subject to friction, it may (though not necessarily) develop a sesamoid bone within its substance. An example is the peroneus longus tendon in the sole of the foot. However, sesamoid bones may also appear in tendons not subject to friction.

## Multiple Attachments

Many muscles have only two attachments, one at each end. However, more complex muscles are often attached to several different structures at their origin and/or their insertion. If these attachments are separated, effectively meaning the muscle gives rise to two or more tendons and/or aponeuroses inserting into different places, the muscle is said to have two or more heads. For example, the biceps brachii has two heads at its origin: one from the coracoid process of the scapula and the other from the supraglenoid tubercle (see p. 178). The triceps brachii has three heads and the quadriceps has four.

## Red and White Muscle Fibers

Historically, three types of skeletal muscle fiber have been distinguished: (1) red slow-twitch fibers, or type I; (2) white fast-twitch fibers, or type IIa; and (3) intermediate fast-twitch fibers, or type IIb. More recently, however, types Ic, IIc, IIac, IIab, type IIm, and others (e.g. type II X) have been described.

1. **Red slow-twitch fibers (type I)**: these fibers are thin cells that contract slowly. The red color is due to their content of myoglobin, a substance similar to hemoglobin, which stores oxygen and increases the rate of oxygen diffusion within the muscle fiber. As long as oxygen supply is plentiful, red fibers can contract for sustained periods, and are thus very resistant to fatigue. Successful marathon runners tend to have a high percentage of these red fibers.

2. **White fast-twitch fibers (type IIa)**: these fibers are large cells that contract rapidly. They are pale because of their lower content of myoglobin. White fibers fatigue quickly, because they rely on short-lived glycogen reserves in the fiber to contract. However, they are capable of generating much more powerful contractions than red fibers, enabling them to perform rapid, powerful movements for short periods. Successful sprinters have a higher proportion of these white fibers.

3. **Intermediate fast-twitch fibers (Type IIb)**: these red or pink fibers are a compromise in size and activity between the red and white fibers.

Note: There is always a mixture of these types of muscle fiber in any given muscle, giving it a range of fatigue resistance and contractile speeds.

## Blood Supply

In general, each muscle receives an arterial supply to bring nutrients via the blood into the muscle, and contains several veins to take away metabolic by-products surrendered by the muscle into the blood. These blood vessels usually enter through the central part of the muscle, but may also enter toward one end. Thereafter, they branch into a capillary plexus, which spreads throughout the intermuscular septa, to eventually penetrate the endomysium around each muscle fiber. During exercise, the capillaries dilate, increasing the amount of blood flow in the muscle by up to 800 times. However, a muscle tendon, because it is composed of a relatively inactive tissue, has a much less extensive blood supply.

## Nerve Supply

The nerve supply to a muscle usually enters at the same place as the blood supply (neurovascular bundle), and branches through the connective tissue septa into the endomysium in a similar way. Each skeletal muscle fiber is supplied by a single nerve ending. This is in contrast to some other muscle tissues, which are able to contract without any nerve stimulation.

The nerve entering the muscle usually contains roughly equal proportions of sensory and motor nerve fibers, although some muscles may receive separate sensory branches. As the nerve fiber approaches the muscle fiber, it divides into a number of terminal branches, collectively called a *motor end plate*.

## Motor Unit of a Skeletal Muscle

A motor unit consists of a single motor nerve cell and the muscle fibers stimulated by it. The motor units vary in size, ranging from cylinders of muscle 5–7mm in diameter in the upper limb to 7–10mm in diameter in the lower limb. The average number of muscle fibers within a unit is 150 (but this number ranges from less than ten to several hundred). Where fine gradations of movement are required, as demanded of the muscles of the eyeball or fingers, the number of muscle fibers supplied by a single nerve cell is small. On the other hand, where grosser movements are required, as demanded of the muscles of the lower limb, each nerve cell may supply a motor unit of several hundred fibers.

The muscle fibers in a single motor unit are spread throughout the muscle, rather than being clustered together. This means that stimulation of a single motor unit will cause the entire muscle to exhibit a weak contraction.

Skeletal muscles work on an "all or nothing principle": in other words, groups of muscle cells, or fasciculi, can either contract or not contract. Depending on the strength of contraction required, a certain number of muscle cells will fully contract, while others will not contract at all. When a greater muscular effort is needed, most of the motor units may be stimulated at the same time. However, under normal conditions, the motor units tend to work in relays, so that during prolonged contractions some are inhibited while others are contracting—this is known as *gradual increments of contraction (GIC)*.

## Muscle Reflexes

Within skeletal muscles there are two specialized types of nerve receptor that can sense tension (length or stretch): muscle spindles and Golgi tendon organs (GTOs). *Muscle spindles* are cigar-like in shape and consist of tiny modified muscle fibers called *intrafusal fibers*, and nerve endings, encased together within a connective tissue sheath; they lie between and parallel to the main muscle fibers. *GTOs* are located mostly at the junctions of muscles and their tendons or aponeuroses.

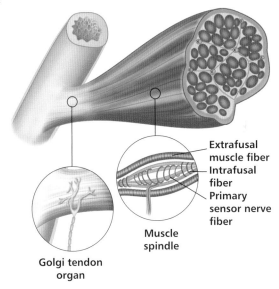

Extrafusal muscle fiber
Intrafusal fiber
Primary sensor nerve fiber
Muscle spindle
Golgi tendon organ

*Figure 2.6. Anatomy of the muscle spindle and Golgi tendon organ.*

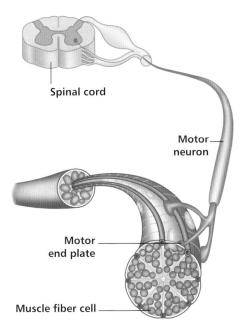

Spinal cord

Motor neuron

Motor end plate

Muscle fiber cell

*Figure 2.5. A motor unit of a skeletal muscle.*

## Stretch Reflex (Mono-synaptic Reflex Arc)

The *stretch reflex* helps control posture by maintaining muscle tone. It also helps prevent injury, by enabling a muscle to respond to a sudden or unexpected increase in length. The way it works is as follows:

1. When a muscle is lengthened, the muscle spindles are excited, causing each spindle to send a nerve impulse communicating the speed of lengthening to the spinal cord.

2. On receiving this impulse, the spinal cord immediately sends a proportionate impulse back to the stretched muscle fibers, causing them to contract, in order to decelerate the movement. This circular process is known as a *reflex arc*.

3. An impulse is simultaneously sent from the spinal cord to the antagonist of the contracting muscle (i.e. the muscle opposing the contraction), causing the antagonist to inhibit, so that it cannot resist the contraction of the stretched muscle. This process is known as *reciprocal inhibition*.

4. Concurrent with this spinal reflex, nerve impulses are also sent up the spinal cord to the brain to relay information about muscle length and the speed of muscle contraction. A reflex in the brain feeds nerve impulses back to the muscles in order to ensure the appropriate muscle tone is maintained to meet the requirements of posture and movement.

5. Meanwhile, the stretch sensitivity of the minute intrafusal muscle fibers within the muscle spindle are smoothed and regulated by gamma efferent nerve fibers*, arising from motor neurons within the spinal cord. Thus, a gamma motor neuron reflex arc ensures the evenness of muscle contraction, which would otherwise be jerky if muscle tone relied on the stretch reflex alone.

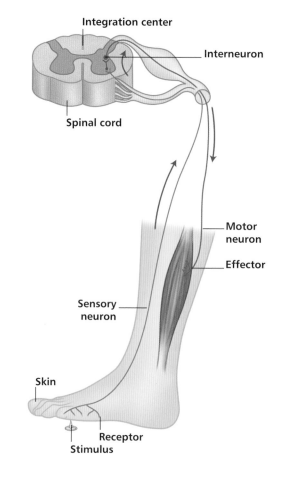

*Figure 2.7: The stretch (monosynaptic reflex arc).*

The classic clinical example of the stretch reflex in action is the knee jerk, or patellar reflex, whereby the patellar tendon is lightly struck with a small rubber hammer. This results in the following sequence of events:

1. The sudden stretch of the patellar tendon in series causes the quadriceps to be stretched, i.e. the sharp tap on the patellar tendon causes a sudden stretch of the tendon.
2. This rapid stretch is registered by the muscle spindles within the quadriceps, causing the quadriceps to contract. This causes a small kick as the knee straightens suddenly, and takes the tension off the muscle spindles.
3. Simultaneously, nerve impulses to the hamstrings (which are the antagonists of the quadriceps) cause functional inhibition of the hamstring muscles.

* The function of these nerve fibers is to regulate the sensitivity of the spindle and the total tension in the muscle.

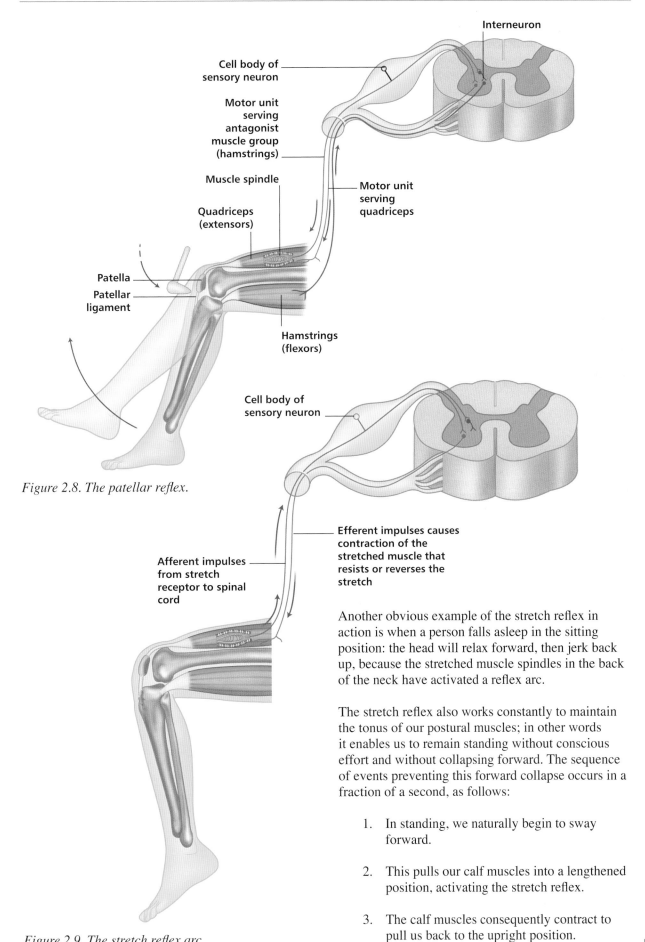

**Interneuron**

Cell body of sensory neuron

Motor unit serving antagonist muscle group (hamstrings)

Muscle spindle

Quadriceps (extensors)

Motor unit serving quadriceps

Patella

Patellar ligament

Hamstrings (flexors)

*Figure 2.8. The patellar reflex.*

Cell body of sensory neuron

Efferent impulses causes contraction of the stretched muscle that resists or reverses the stretch

Afferent impulses from stretch receptor to spinal cord

Another obvious example of the stretch reflex in action is when a person falls asleep in the sitting position: the head will relax forward, then jerk back up, because the stretched muscle spindles in the back of the neck have activated a reflex arc.

The stretch reflex also works constantly to maintain the tonus of our postural muscles; in other words it enables us to remain standing without conscious effort and without collapsing forward. The sequence of events preventing this forward collapse occurs in a fraction of a second, as follows:

1. In standing, we naturally begin to sway forward.

2. This pulls our calf muscles into a lengthened position, activating the stretch reflex.

3. The calf muscles consequently contract to pull us back to the upright position.

*Figure 2.9. The stretch reflex arc.*

## Deep Tendon Reflex (Autogenic Inhibition)

In contrast to the stretch reflex, which involves the muscle spindle's response to muscle elongation, the *deep tendon reflex* involves the reaction of GTOs to muscle contraction or undue rise in tension. Accordingly, the deep tendon reflex creates the opposite effect to that of the stretch reflex. The way it works is as follows:

1. When a muscle contracts, it pulls on the tendons situated at either end of the muscle.

2. The tension in the tendon causes the GTOs to transmit impulses to the spinal cord (some impulses continue to the cerebellum).

3. As these impulses reach the spinal cord, they inhibit the motor nerves supplying the contracting muscle, thus reducing tonus.

4. Simultaneously, the motor nerves supplying the antagonist muscle are activated, causing it to contract. This process is called *reciprocal activation*.

5. Meanwhile, the information reaching the cerebellum is processed and fed back to help readjust muscle tension.

The deep tendon reflex has a protective function: it prevents the muscle from contracting so hard that it would rip its attachment off the bone. It is therefore especially important during activities which involve rapid switching between flexion and extension, such as running.

Note, however, that in normal day-to-day movement, tension in the muscles is not sufficient to activate the GTOs' deep tendon reflex. By contrast, the threshold of the muscle spindle stretch reflex is set much lower, because it must constantly maintain sufficient tonus in the postural muscles to keep the body upright.

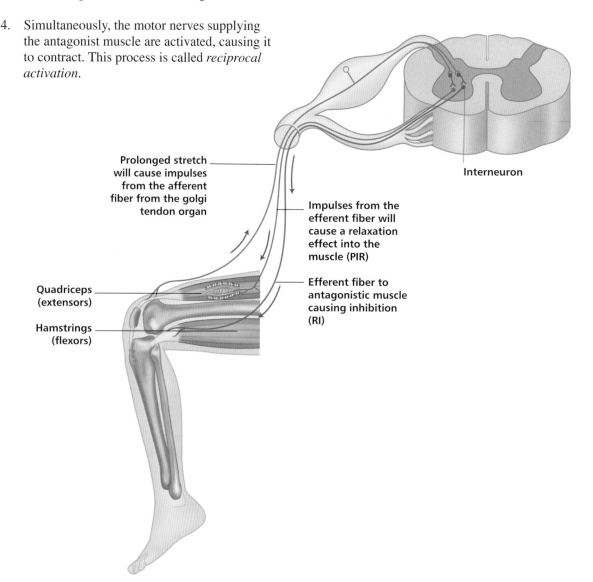

Interneuron

Prolonged stretch will cause impulses from the afferent fiber from the golgi tendon organ

Impulses from the efferent fiber will cause a relaxation effect into the muscle (PIR)

Quadriceps (extensors)

Efferent fiber to antagonistic muscle causing inhibition (RI)

Hamstrings (flexors)

*Figure 2.10. The deep tendon reflex.*

## Muscle Contractions

A muscle will contract upon stimulation, in an attempt to bring its attachments closer together, but this does not necessarily result in a shortening of the muscle. If the contraction of a muscle results in no movement, such a contraction is called *isometric*; if movement of some sort results, the contraction is called *isotonic*.

### Isometric Contraction

An *isometric* contraction occurs when a muscle increases its tension, but the length of the muscle is not altered. In other words, although the muscle tenses, the joint over which the muscle works does not move. One example of this is in the hand with the elbow held stationary and bent. Trying to lift something that proves to be too heavy to move is another example. Note also that some of the postural muscles are largely working isometrically by automatic reflex. For example, in the upright position, the body has a natural tendency to fall forward at the ankle; this is prevented by isometric contraction of the calf muscles. Likewise, the center of gravity of the skull would make the head tilt forward if the muscles at the back of the neck did not contract isometrically to keep the head centralized.

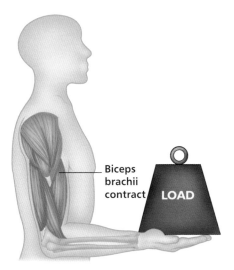

*Figure 2.11. Isometric contraction, for example, holding a heavy object at 90 degrees in a stationary position.*

### Isotonic Contraction

*Isotonic* contractions of muscle enable us to move about. There are of two types of isotonic contraction: concentric and eccentric.

### Concentric Contraction

In *concentric* contractions, the muscle attachments move closer together, causing movement at the joint. Using the example of holding an object, if biceps brachii contracts concentrically, the elbow joint will flex and the hand will move toward the shoulder, against gravity. Similarly, if we do an abdominal crunch on a Swiss ball, the abdominal muscles must contract concentrically to raise the torso (see figure 2.12).

*Figure 2.12. Abdominals contract concentrically to raise the body.*

### Eccentric Contraction

*Eccentric* contraction means that the muscle fibers "pay out" in a controlled manner to slow down movements in a case where gravity, if unchecked, would otherwise cause them to be too rapid—for example, lowering an object held in the hand down to your side. Another example is simply sitting down into a chair. Therefore, the difference between concentric and eccentric contractions is that in the former, the muscle shortens, and in the latter, it actually lengthens.

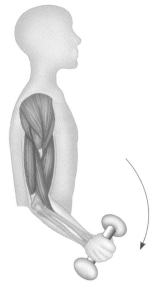

*Figure 2.13. Biceps brachii contracts eccentrically to lower an object (dumbbell) down to the side.*

## Muscle Shapes (Arrangement of Fascicles)

Muscles come in a variety of shapes according to the arrangement of their fascicles. The reason for this variation is to provide optimum mechanical efficiency for a muscle in relation to its position and action. The most common arrangements of fascicles yield muscle shapes which can be described as parallel, pennate, convergent, and circular, with each of these shapes having further subcategories. The different shapes are illustrated in Figure 2.14.

### Parallel

In this arrangement the fascicles run parallel to the long axis of the muscle. If the fascicles extend throughout the length of the muscle, it is known as a *strap* muscle—for example, the sartorius. If the muscle also has an expanded belly and tendons at both ends, it is called a *fusiform* muscle—for example, the biceps brachii. A variation of this type of muscle has a fleshy belly at either end, with a tendon in the middle; such muscles are referred to as *digastric*.

### Pennate

*Pennate* muscles are so named because their short fasciculi are attached obliquely to the tendon, like the structure of a feather (Latin *penna* = "feather"). If the tendon develops on one side of the muscle, it is referred to as *unipennate*—for example, the flexor digitorum longus in the leg. If the tendon is in the middle and the fibers are attached obliquely from both sides, it is known as *bipennate*, a good example of which is the rectus femoris. If there are numerous tendinous intrusions into the muscle, with fibers attaching obliquely from several directions (thus resembling many feathers side by side), the muscle is referred to as *multipennate*; the best example is the middle part of the deltoid muscle.

### Convergent

Muscles that have a broad origin with fascicles converging toward a single tendon, giving the muscle a triangular shape, are called *convergent* muscles. The best example is the pectoralis major.

### Circular

When the fascicles of a muscle are arranged in concentric rings, the muscle is referred to as *circular*. All the sphincter skeletal muscles in the body are of this type; they surround openings, which they close by contracting. An example is the orbicularis oculi.

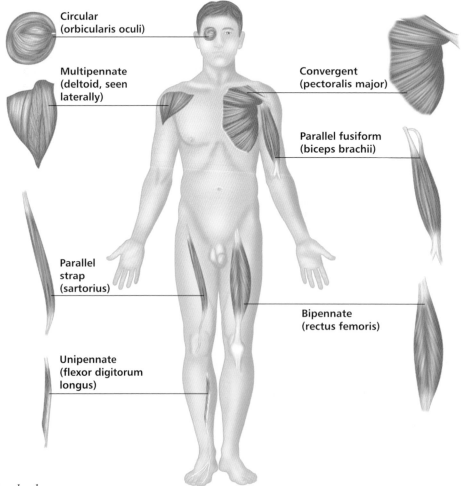

Circular
(orbicularis oculi)

Multipennate
(deltoid, seen
laterally)

Convergent
(pectoralis major)

Parallel fusiform
(biceps brachii)

Parallel
strap
(sartorius)

Bipennate
(rectus femoris)

Unipennate
(flexor digitorum
longus)

*Figure 2.14. Muscle shapes.*

## Range of Motion Versus Power

When a muscle contracts, it can shorten by up to 70% of its original length; hence, the longer the fibers, the greater the range of motion. On the other hand, the strength of a muscle depends on the total number of muscle fibers it contains, rather than the length of the fibers. Therefore:

1. Muscles with long parallel fibers produce the greatest range of motion, but are not usually very powerful.

2. Muscles with a pennate pattern (especially multipennate) pack in the most fibers. Such muscles shorten less than long parallel muscles, but tend to be much more powerful.

### Functional Characteristics of a Skeletal Muscle

Everything that has been said about muscles so far in this book enables us to formulate a list of functional characteristics pertaining to skeletal muscle.

## Excitability

*Excitability* is the ability to receive and to respond to a stimulus. In the case of a muscle, when a nervous impulse from the brain reaches the muscle, a chemical known as *acetylcholine* is released. This chemical produces a change in the electrical balance in the muscle fiber and, as a result, generates an electrical current known as an *action potential*. The action potential conducts the electrical current from one end of the muscle cell to the other and results in a contraction of the muscle cell, or muscle fiber (remember that one muscle cell = one muscle fiber).

## Contractility

*Contractility* is the ability of a muscle to shorten forcibly when stimulated. The muscles themselves can only contract; they cannot lengthen, except via some external means (i.e. manually), beyond their normal resting length (see "Tonus" below). In other words, muscles can only pull their ends together (contract)—they cannot push them apart.

## Extensibility

*Extensibility* is the ability of a muscle to be extended, or returned to its resting length (which is a semi-contracted state) or slightly beyond. For example, if we bend forward at the hips from standing, the muscles of the back, such as the erector spinae, lengthen eccentrically to lower the trunk, paying out slightly beyond their normal resting length, and are thus effectively "elongated."

## Elasticity

*Elasticity* describes the ability of a muscle fiber to recoil after being lengthened, and therefore resume its resting length when relaxed. In a whole muscle, the elastic effect is supplemented by the important elastic properties of the connective tissue sheaths (endomysium and epimysium). Tendons also contribute some elastic properties. An example of this elastic recoil effect can be experienced when coming back up from a forward bend at the hips as described above. Initially there is no muscle contraction; instead, the upward movement is initiated purely by elastic recoil of the back muscles, after which the contraction of the back muscles completes the movement.

## Tonus

*Tonus*, or *muscle tone*, is the term used to describe the slightly contracted state to which muscles return during the resting state. Muscle tonus does not produce active movements, but it keeps the muscles firm, healthy, and ready to respond to stimulation. It is the tonus of skeletal muscles that also helps stabilize and maintain posture. *Hypertonic* muscles are those muscles whose "normal" resting state is over-contracted.

### General Functions of Skeletal Muscles

- **Enable movement**: skeletal muscles are responsible for all locomotion and manipulation, and they enable you to respond quickly.

- **Maintain posture**: skeletal muscles support an upright posture against the pull of gravity.

- **Stabilize joints**: skeletal muscles and their tendons stabilize joints.

- **Generate heat**: skeletal muscles (in common with smooth and cardiac muscles) generate heat, which is important in maintaining a normal body temperature.

See Appendix 2 for the main muscles involved in different movements of the body.

## The Skeletal System

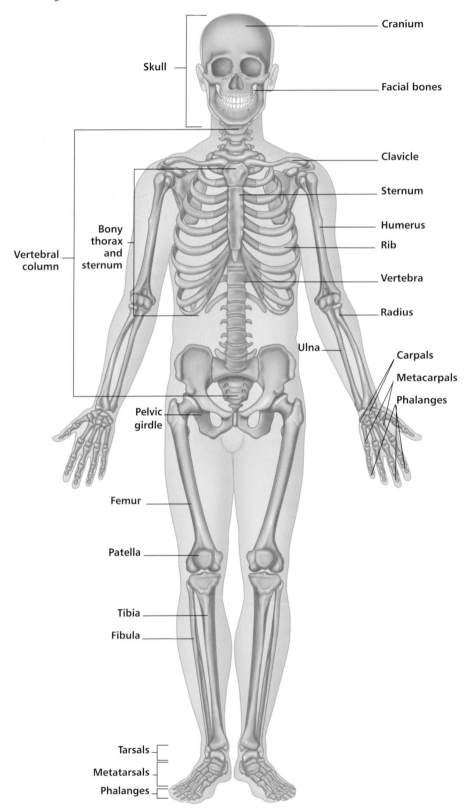

Figure 2.15. (a) Skeleton: anterior view.

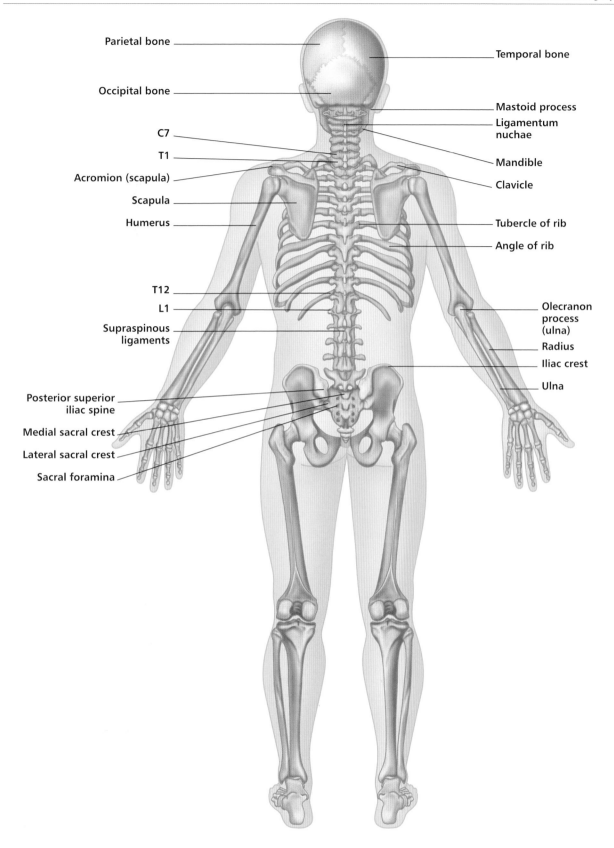

Parietal bone

Temporal bone

Occipital bone

Mastoid process

Ligamentum nuchae

C7

T1

Mandible

Acromion (scapula)

Clavicle

Scapula

Humerus

Tubercle of rib

Angle of rib

T12

L1

Olecranon process (ulna)

Supraspinous ligaments

Radius

Iliac crest

Ulna

Posterior superior iliac spine

Medial sacral crest

Lateral sacral crest

Sacral foramina

*Figure 2.15. (b) Skeleton: posterior view.*

## Sections of the Vertebral Column

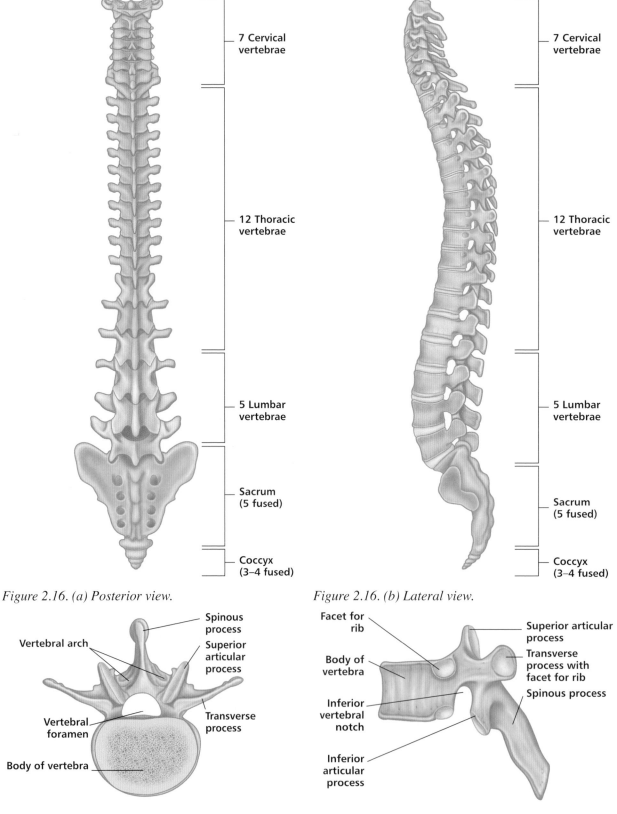

7 Cervical
vertebrae

12 Thoracic
vertebrae

5 Lumbar
vertebrae

Sacrum
(5 fused)

Coccyx
(3–4 fused)

*Figure 2.16. (a) Posterior view.*

7 Cervical
vertebrae

12 Thoracic
vertebrae

5 Lumbar
vertebrae

Sacrum
(5 fused)

Coccyx
(3–4 fused)

*Figure 2.16. (b) Lateral view.*

Spinous
process

Superior
articular
process

Vertebral arch

Transverse
process

Vertebral
foramen

Body of vertebra

Facet for
rib

Body of
vertebra

Inferior
vertebral
notch

Inferior
articular
process

Superior articular
process

Transverse
process with
facet for rib

Spinous process

*Figure 2.16. (c) Vertebrae: lumbar (superior view) and thoracic (lateral view).*

## Thoracic to Pelvic Region

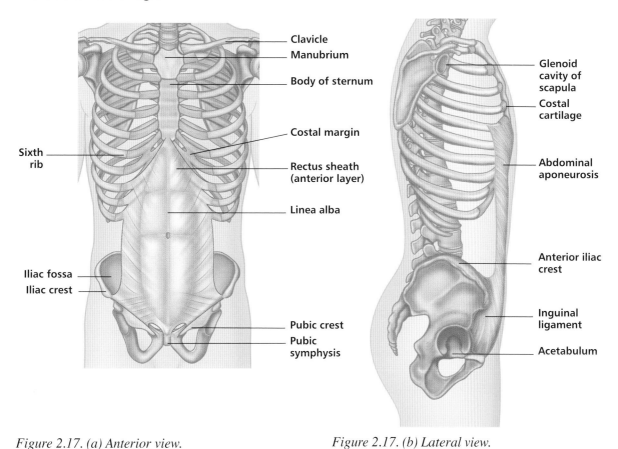

Figure 2.17. (a) Anterior view.

Figure 2.17. (b) Lateral view.

## Scapula

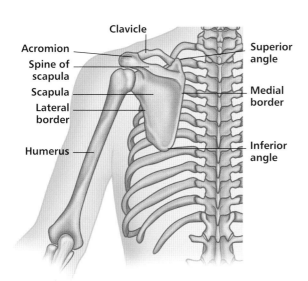

Figure 2.18. Posterior view.

## Skull to Sternum

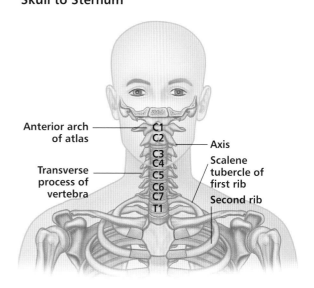

Figure 2.19. Anterior view (the mandible and maxilla are removed).

## Skull to Humerus

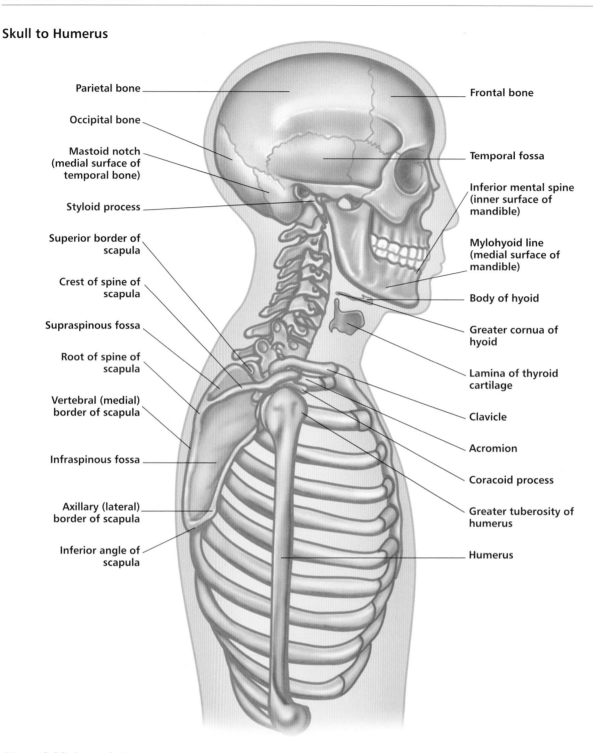

Parietal bone

Occipital bone

Mastoid notch
(medial surface of
temporal bone)

Styloid process

Superior border of
scapula

Crest of spine of
scapula

Supraspinous fossa

Root of spine of
scapula

Vertebral (medial)
border of scapula

Infraspinous fossa

Axillary (lateral)
border of scapula

Inferior angle of
scapula

Frontal bone

Temporal fossa

Inferior mental spine
(inner surface of
mandible)

Mylohyoid line
(medial surface of
mandible)

Body of hyoid

Greater cornua of
hyoid

Lamina of thyroid
cartilage

Clavicle

Acromion

Coracoid process

Greater tuberosity of
humerus

Humerus

*Figure 2.20. Lateral view.*

# Synovial Joints

Synovial joints possess a joint cavity that contains *synovial fluid*. They are freely movable, or *diarthrotic*, and have a number of distinguishing features:

**Articular cartilage** (or *hyaline cartilage*), which covers the ends of the bones that form the joint.

A **joint cavity**, which is more a potential space than a real one, because it is filled with lubricating *synovial fluid*. The joint cavity is enclosed by a double-layered "sleeve", or capsule, known as the *articular capsule*. The external layer of the articular capsule is known as the *capsular ligament*: this is a tough, flexible, fibrous connective tissue that is continuous with the periostea of the articulating bones. The internal layer, or *synovial membrane*, is a smooth membrane made of loose connective tissue that lines the capsule and all internal joint surfaces other than those covered in hyaline cartilage.

**Synovial fluid**, which occupies the free spaces within the joint capsule. This slippery fluid is also found within the articular cartilage and provides a film that reduces friction between the cartilages. When a joint is compressed by movement, synovial fluid is forced out of the articular cartilage; when pressure is relieved, the fluid rushes back in. Synovial fluid nourishes the cartilage, which has a lower vascularity; it also contains *phagocytic cells* (cells that eat dead matter), which rid the joint cavity of microbes or cellular waste. The amount of synovial fluid varies in different joints, but is always sufficient to form a thin film to reduce friction. When a joint is injured, extra fluid is produced and creates the characteristic swelling of the joint. The synovial membrane later reabsorbs this extra fluid.

**Collateral or accessory ligaments**, which determine the strength and movement of a joint. Synovial joints are reinforced and strengthened by a number of ligaments: these are either *capsular*, i.e. thickened parts of the fibrous capsule itself, or independent *collateral* ligaments, which are distinct from the capsule. Depending on their position and quantity around the joint, ligaments will restrict movement in certain directions, as well as preventing unwanted movement. Ligaments always bind bone to bone, and generally the more ligaments a joint has, the stronger it is.

**Bursae**, which are fluid-filled sacs often found cushioning the joint. They are lined by a synovial membrane and contain synovial fluid. Bursae are found between tendons and bone, ligament and bone, or muscle and bone, and reduce friction by acting as a cushion.

**Tendon sheaths**, which wrap themselves around tendons subject to friction, in order to protect them. They are frequently found in close proximity to synovial joints and have the same structure as bursae.

**Articular discs (menisci)**, which act as shock absorbers (similar to the fibrocartilaginous disc in the pubic symphysis) and are present in some synovial joints. For example, in the knee joint two crescent-shaped fibrocartilage discs, called the *medial* and *lateral menisci*, lie between the medial and lateral condyles of the femur and the medial and lateral condyles of the tibia.

## Notes About Synovial Joints

- Some tendons run partly within the joint and are therefore *intracapsular*.
- The fibers of many ligaments are largely integrated with those of the capsule, and the delineation between capsule and ligament is sometimes unclear.
- Ligaments are termed *intracapsular* (or *intra-articular*) when inside the joint cavity, and *extracapsular* (or *extra-articular*) when outside the capsule.
- Many ligaments of the knee joint are modified extensions or expansions of muscle tendons, but are classed as ligaments in order to differentiate them from the more regular stabilizing tendons. An example is the patellar ligament from the quadriceps.
- Most synovial joints have various bursae in their vicinity pertaining to each joint.

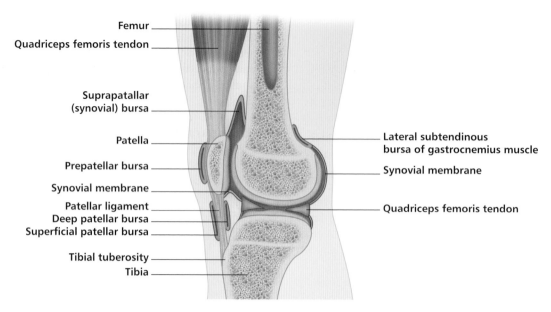

*Figure 2.21. A typical synovial joint: the knee joint (mid-sagittal view).*

Labels: Femur, Quadriceps femoris tendon, Suprapatallar (synovial) bursa, Patella, Prepatellar bursa, Synovial membrane, Patellar ligament, Deep patellar bursa, Superficial patellar bursa, Tibial tuberosity, Tibia, Lateral subtendinous bursa of gastrocnemius muscle, Synovial membrane, Quadriceps femoris tendon

## Subclassifications of Synovial (Diarthrotic) Joints

There are several types of synovial joint: plane (or gliding), hinge, pivot, ball-and-socket, condyloid, saddle, and ellipsoid. An example of each synovial joint is highlighted in bold.

### Plane (or Gliding)

In *plane* joints, movement occurs when two, generally flat or slightly curved, surfaces glide across one another. Examples: the acromioclavicular joint, the joints between the carpal bones in the wrist or between the tarsal bones in the ankle, the **facet joints** between the vertebrae, and the sacroiliac joint.

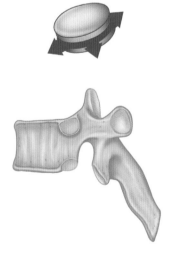

### Hinge

In *hinge* joints, movement occurs around only one axis, a transverse one, as in the hinge of the lid of a box. A protrusion of one bone fits into a concave or cylindrical articular surface of another, permitting flexion and extension. Examples: the interphalangeal joints, the **elbow**, and the knee.

### Pivot

In *pivot* joints, movement takes place around a vertical axis, like the hinge of a gate. A more or less cylindrical articular surface of bone protrudes into and rotates within a ring formed by bone or ligament. Examples: the dens of the axis protrude through the hole in the atlas, allowing the rotation of the head from side to side; also, the **joint** between the **radius** and the **ulna** at the elbow allows the round head of the radius to rotate within a "ring" of ligament that is secured to the ulna.

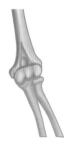

### Ball and Socket

*Ball and socket* joints consist of a "ball" formed by the spherical or hemispherical head of one bone that rotates within the concave "socket" of another, allowing flexion, extension, adduction, abduction, circumduction, and rotation. Thus, they are multiaxial and allow the greatest range of motion of all joints. Examples: the **shoulder joint** and the hip joint

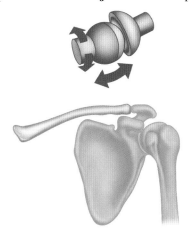

### Condyloid

In common with ball and socket joints, *condyloid* joints have a spherical articular surface that fits into a matching concavity. In addition, like ball and socket joints, condyloid joints permit flexion, extension, abduction, adduction, and circumduction. However, the disposition of surrounding ligaments and muscles prevents active rotation around a vertical axis. Examples: the **metacarpophalangeal joints** of the fingers (but not the thumb).

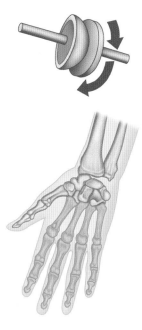

### Saddle

*Saddle* joints are similar to condyloid joints, except that the two articulating surfaces respectively have convex and concave areas, which fit together like a saddle and a horse's back. These joints permit even more movement than condyloid joints. Example: the **carpometacarpal joint** of the thumb, which allows opposition of the thumb to the fingers.

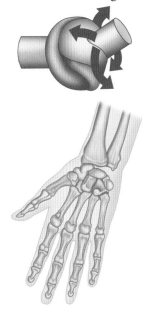

### Ellipsoid

An *ellipsoid* joint is effectively a ball and socket joint, but the articular surfaces are ellipsoidal instead of spherical. The movements allowed by these joints are similar to those of ball and socket joints, with the exception of rotation, which is prevented because of the shape of the ellipsoidal surfaces. Example: the **radiocarpal joint.**

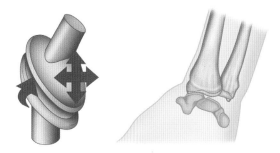

*Figure 2.22. The different types of synovial joints, with an example of each joint given.*

# Musculoskeletal Mechanics

### Origins and Insertions

At the beginning of this chapter we briefly discussed the origins and insertions of muscles. In the majority of movements, one attachment of a muscle remains relatively stationary while the attachment at the other end moves. The more stationary attachment is usually referred to as the *origin* of the muscle, and the other attachment, the *insertion*. A spring that closes a gate could be said to have its origin on the gatepost and its insertion on the gate itself.

In the body, however, the arrangement is rarely so clear-cut, because, depending on the activity one is engaged in, the fixed and movable ends of the muscle may be reversed. For example, muscles that attach the upper limb to the chest normally move the arm relative to the trunk, which means their origins are on the trunk and their insertions are on the upper limb. However, in climbing, the arms are fixed, while the trunk moves as it is pulled up to the fixed limbs. In this type of situation, where the insertion is fixed and the origin moves, the muscle is said to perform a *reversed action*. Because there are so many situations where muscles are working with a reversed action, it is sometimes less confusing to simply speak of *attachments*, without reference to origin and insertion.

In practice, a muscle attachment that lies more proximally (more toward the trunk or on the trunk) is usually referred to as the *origin*. An attachment that lies more distally (away from the attached end of a limb or away from the trunk) is referred to as the *insertion*.

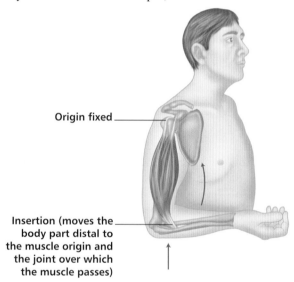

Origin fixed

Insertion (moves the body part distal to the muscle origin and the joint over which the muscle passes)

*Figure 2.23. A muscle working with its origin fixed and its insertion moving.*

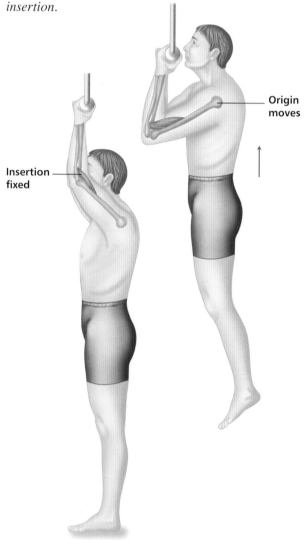

Origin moves

Insertion fixed

*Figure 2.24. Climbing: the muscles are working with the insertion fixed and the origin moving (reversed action).*

## Group Action of Muscles

Muscles work together or in opposition to achieve a wide variety of movements; therefore, whatever one muscle can do, there is another muscle that can undo it. Muscles may also be required to provide additional support or stability to enable certain movements to occur elsewhere.

Muscles are classified into four functional groups:

1. Prime mover, or agonist
2. Antagonist
3. Synergist
4. Fixator

### Prime Mover, or Agonist

A *prime mover* (also called an *agonist*) is a muscle that contracts to produce a specific movement. An example is the biceps brachii, which is the prime mover in elbow flexion. Other muscles may assist the prime mover in providing the same movement, albeit with less effect: such muscles are called *assistant* or *secondary* movers. For example, the brachialis assists the biceps brachii in flexing the elbow, and is therefore a secondary mover.

### Antagonist

The muscle on the opposite side of a joint to the prime mover, and which must relax to allow the prime mover to contract, is called an *antagonist*. For example, when the biceps brachii on the front of the arm contracts to flex the elbow, the triceps brachii on the back of the arm must relax to allow this movement to occur. When the movement is reversed (i.e. the elbow is extended), the triceps brachii becomes the prime mover and the biceps brachii assumes the role of antagonist.

### Synergist

*Synergists* prevent any unwanted movements that might occur as the prime mover contracts. This is especially important where a prime mover crosses two joints, because when it contracts, it will cause movement at both joints, unless other muscles act to stabilize one of the joints. For example, the muscles that flex the fingers not only cross the finger joints, but also cross the wrist joint, potentially causing movement at both joints. However, because you have other muscles acting synergistically to stabilize the wrist joint, you are able to flex the fingers into a fist without also flexing the wrist at the same time.

A prime mover may have more than one action, so synergists also act to eliminate the unwanted movements. For example, the biceps brachii will flex the elbow, but its line of pull will also supinate the forearm (twist the forearm, as in tightening a screw).

If you want flexion to occur without supination, other muscles must contract to prevent this supination. In this context, such synergists are sometimes called *neutralizers*.

### Fixator

A synergist is more specifically referred to as a *fixator* or *stabilizer* when it immobilizes the bone of the prime mover's origin, thus providing a stable base for the action of the prime mover. The muscles that stabilize (fix) the scapula during movements of the upper limb are good examples. The sit-up exercise gives another good example. The abdominal muscles attach to both the ribcage and the pelvis. When they contract to enable you to perform a sit-up, the hip flexors will contract synergistically as fixators to prevent the abdominals tilting the pelvis, thereby enabling the upper body to curl forward as the pelvis remains stationary.

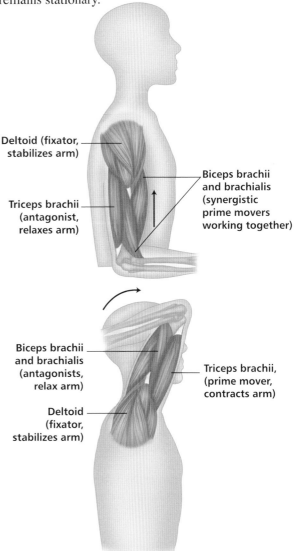

*Figure 2.25. Group action of muscles: (a) flexing the arm at the elbow; (b) extending the arm at the elbow (showing reversed roles of prime mover and antagonist).*

## Leverage

In classical biomechanics, the bones, joints, and muscles together form a system of levers in the body, to optimize the relative strength, range, and speed required of any given movement. The joints act as fulcrums, the muscles apply the efforts, and the bones bear the weight of the body part to be moved.

A muscle attached close to the fulcrum will be weaker than it would be if it were attached further away. However, it is able to produce a greater range and speed of movement, because the length of the lever amplifies the distance traveled by its movable attachment. Figure 2.26 illustrates this principle in relation to the adductors of the hip joint. The muscle so positioned to move the greater load (in this case, the adductor longus) is said to have a *mechanical advantage*. The muscle attached close to the fulcrum is said to operate at a *mechanical disadvantage*, although it can move a load more rapidly through larger distance.

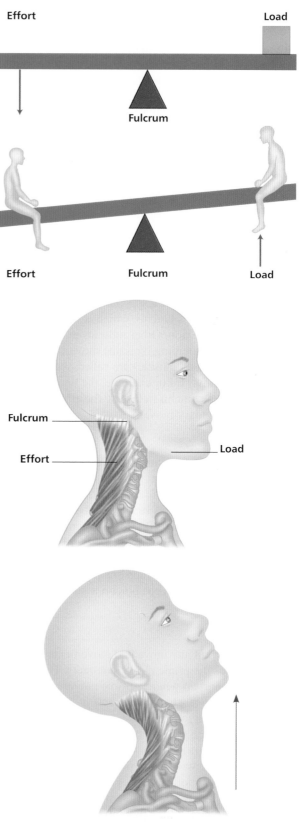

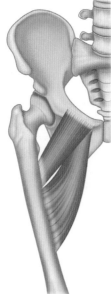

*Figure 2.26. The pectineus is attached closer to the axis of movement than the adductor longus. Therefore, the pectineus is the weaker adductor of the hip, but is able to produce a greater movement of the lower limb per centimeter of contraction.*

Figures 2.27–2.29 illustrate the differences in first-, second-, and third-class levers, with examples in the human body.

*Figure 2.27. First-class lever: the relative position of the components is load–fulcrum–effort. Examples are a seesaw and a pair of scissors. In the body, an example is the ability to extend the head and neck: here the facial structures are the load, the atlanto-occipital joint is the fulcrum, and the posterior neck muscles provide the effort.*

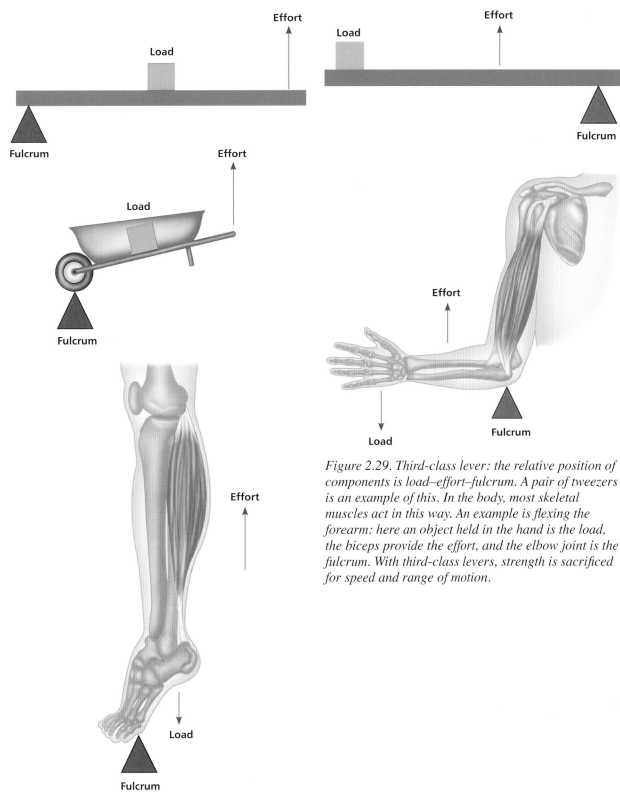

Figure 2.29. *Third-class lever: the relative position of components is load–effort–fulcrum. A pair of tweezers is an example of this. In the body, most skeletal muscles act in this way. An example is flexing the forearm: here an object held in the hand is the load, the biceps provide the effort, and the elbow joint is the fulcrum. With third-class levers, strength is sacrificed for speed and range of motion.*

Figure 2.28. *Second-class lever: the relative position of components is fulcrum–load–effort. The best example is a wheelbarrow. In the body, an example is the ability to raise the heels off the ground in standing: here the ball of the foot is the fulcrum, the body weight is the load, and the calf muscles provide the effort. With second-class levers, speed and range of motion are sacrificed for strength.*

## Muscle Factors That Limit Skeletal Movement

The inability of a muscle to contract or lengthen beyond a certain point can cause some practical hindrances to body movement: these are outlined below.

### Passive Insufficiency

Muscles that span two joints are said to be *biarticular*. These muscles may be unable to "pay out" sufficiently to allow full movement of both joints simultaneously, unless the muscle has been trained to relax. For example, most people need to bend their knees in order to touch their toes; this is because the hamstrings (which span the hip and knee joints) cannot lengthen enough to allow full flexion at the hip joint without also pulling the knee joint into flexion. For the same reason, it is easier to pull your thigh to your chest if your knee is bent than it is with your knee straight. This limitation is called *passive insufficiency*. Passive insufficiency is therefore the inability of a muscle to lengthen by more than a fixed percentage of its length.

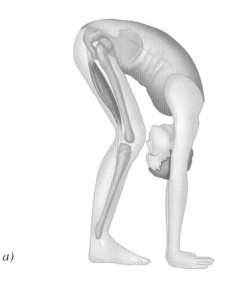
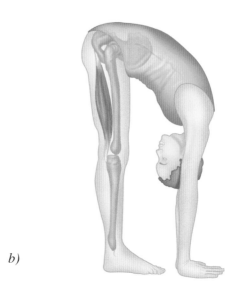

a)

b)

*Figure 2.30. Passive insufficiency example 1: (a) having to bend the knees to the touch the toes means there is passive insufficiency of the hamstrings; (b) being able to touch the toes with the knees straight means there is much less passive insufficiency of the hamstrings.*

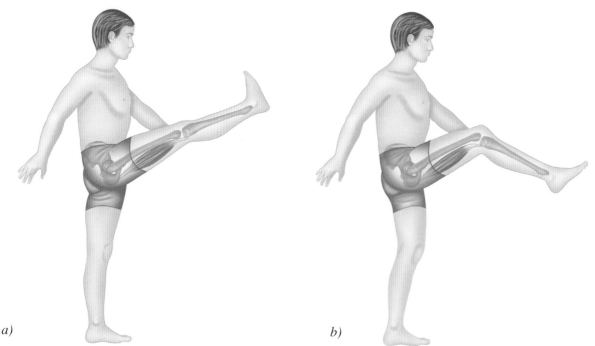

a)

b)

*Figure 2.31. Passive insufficiency example 2: (a) a high kick with the knee straight is possible only if the hamstrings have been trained to overcome their passive insufficiency; (b) for most people, an attempt to perform a high kick will be restricted by hamstring passive insufficiency, causing the knee to bend.*

## Active Insufficiency

*Active insufficiency* is the opposite of passive insufficiency. Whereas passive insufficiency results from the inability of a muscle to lengthen by more than a fixed percentage of its length, active insufficiency is the inability of a muscle to contract by more than a fixed amount. For example, most people can flex their knee to bring their heel close to their buttock when their hip is flexed, because the upper part of the hamstrings is lengthened and the lower part is shortened. However, one is normally unable to fully flex the knee when the hip is extended; this is because with the hip extended, the hamstrings are already shortened, meaning that there is insufficient "shortening" potential remaining in the hamstrings to then fully flex the knee.

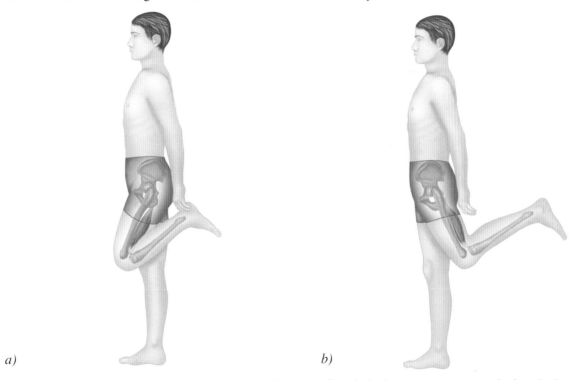

a)

b)

*Figure 2.32. Active insufficiency example: (a) with the hip flexed, the hamstrings are stretched at the hip, enabling their contraction to fully flex the knee; (b) with the hip extended, the shortened hamstrings are unable to contract any further to fully flex the knee.*

## Concurrent Movement

If extension of the hip is required at the same time as extension of the knee, as in the push-off from the ground in running, the phenomenon known as *concurrent movement* applies, and proves very useful. To grasp the concept of concurrent movement, first remember that when the hamstrings contract, they are able to both extend the hip joint and flex the knee joint, either singly or simultaneously. In analyzing the example of running in more detail, we therefore observe the following:

- As the foot pushes against the ground, the hamstrings contract to extend the hip.
- Meanwhile, the fixators prevent the hamstrings from flexing the knee.
- Consequently, the hamstrings are shortened only at their upper end (origin), but remain lengthened at their lower end (insertion).

- The antagonist to the hamstring's action of extending the hip is the rectus femoris, which relaxes because of reciprocal inhibition (see p. 26) to allow the hamstrings to contract.
- When the hip is well extended, the already stretched rectus femoris is unable to lengthen further, causing it to pull the knee into extension.
- The rectus femoris is therefore lengthened at its upper end and shortened at its lower end.

Concurrent movement therefore avoids passive and active insufficiency of the hamstrings and rectus femoris by neither shortening nor stretching both ends of either muscle; rather, one end lengthens as the other shortens, and vice versa in the other muscle. Figure 2.33 should elucidate this concept.

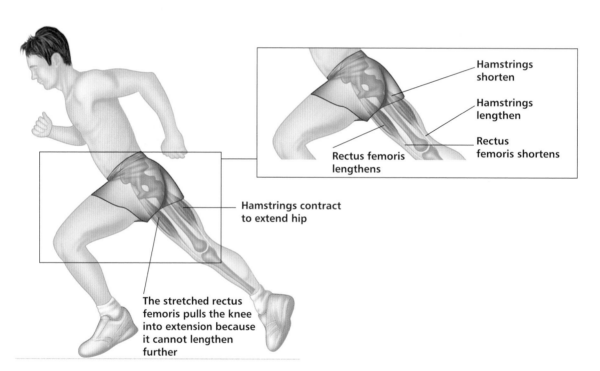

**Hamstrings shorten**

**Hamstrings lengthen**

**Rectus femoris shortens**

**Rectus femoris lengthens**

**Hamstrings contract to extend hip**

**The stretched rectus femoris pulls the knee into extension because it cannot lengthen further**

*Figure 2.33. Concurrent movement.*

## Countercurrent Movement

If flexion of the hip is required to occur at the same time as extending the knee, as in kicking a ball, a *countercurrent movement* occurs. In analyzing the example of kicking in more detail, we therefore observe the following:

- In the action of kicking a ball, the rectus femoris acts as a prime mover to flex the hip and extend the knee.
- Thus, both the upper and lower portions of the rectus femoris are shortened.
- The hamstrings relax because of reciprocal inhibition, so that they can extend at both ends and allow the kick to occur.

- The rectus femoris relaxes once the movement has been made, but the momentum of the movement is still propelling the leg forward.
- At this stage, the hamstrings contract to act as a "brake" for the leg, as it flies forward.

Countercurrent movements therefore prevent injury by ensuring that the antagonist relaxes first, then contracts at the right time to prevent the forces of momentum from overstretching muscles and ligaments. So-called *ballistic* movements rely on this principle, but are often done so forcefully that the power of momentum is greater than the ability of the antagonist to "brake" that momentum. In such instances, muscle and ligament damage often occurs.

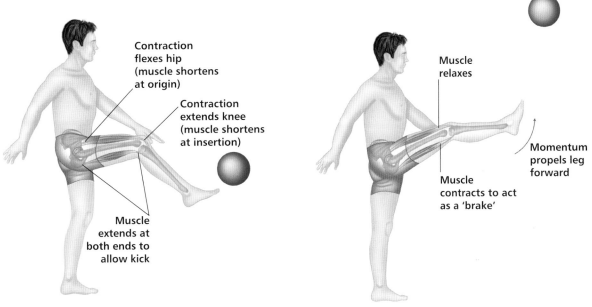

Contraction flexes hip (muscle shortens at origin)

Contraction extends knee (muscle shortens at insertion)

Muscle extends at both ends to allow kick

Muscle relaxes

Momentum propels leg forward

Muscle contracts to act as a 'brake'

*Figure 2.34. Countercurrent movement.*

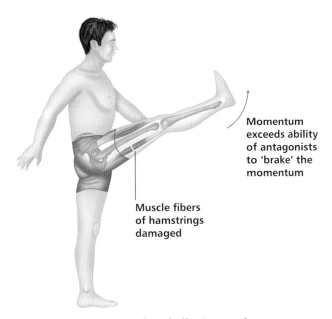

Momentum exceeds ability of antagonists to 'brake' the momentum

Muscle fibers of hamstrings damaged

*Figure 2.35. Damage that can be caused by an overzealous ballistic stretch.*

## Core Stability

During day-to-day activities, skeletal muscles act either as stabilizing muscles or as muscles of movement (as outlined under "Group Action of Muscles"—see p. 41). Stabilizing muscles maintain posture, or hold the body in a given position as a "platform," so that other muscles can cause the body to move in some way.

Stabilizing muscles tend to be situated deep within the body. To maintain posture or a steady platform, their fibers perform a minimal contraction over an extended period of time; hence these muscles are built for endurance and therefore have many slow-twitch fibers (see "Red and White Muscle Fibers," p. 24). Persons with poor postural alignment or an inactive lifestyle tend to have insufficient tone in these muscles, which further exacerbates their poor posture and lessens their ability to stabilize functional movements.

When the stabilizing muscles are underused, the nerve impulses find it more difficult to get through to those muscles, leading to what is referred to as *poor recruitment*. This means that, if we do not use a muscle for an extended period of time, we will find it more difficult to re-innervate that muscle in order to use it again. Consequently, the majority of people in modern society would benefit from exercises that specifically address their neglected deep postural muscles.

It is particularly important to maintain your torso as a stable platform relative to the movements carried out by your limbs. As your torso or mid-section is the "core" of your body, its success as a stable platform is referred to as *core stability*. Good core stability therefore allows you to maintain a rigid mid-section, without gravity or other forces interfering with the movement you wish to perform. Core stability muscles can be retrained, especially through bracing and stabilizing exercises—a fact utilized in physiotherapy treatment, Pilates, Taijiquan, Hatha Yoga, and so on. In essence, core stability can be summarized as the successful recruitment of deep muscles that maintain the natural curvatures (neutral alignment) of the spine during all other movements of the body.

Good core stability results from the deep stabilizing trunk muscles coordinating their contractions to stabilize the spine, rather like the tightening of guide ropes around a pole or mast to give it strength and maintain its position.

The deep stabilizing, or core stability, muscles collectively create what is known as an *inner unit* of muscle. These muscles include the transversus abdominis, multifidus, pelvic floor, diaphragm, and posterior fibers of the internal oblique. The main muscles that initiate movement of the limbs while working in unison with the inner unit are collectively referred to as the *outer unit* or *global muscles*; these are the spinal erectors, external and internal obliques, latissimus dorsi, glutes, hamstrings, and adductors. The following effects summarize how core stability is enhanced by some subsidiary factors of body mechanics.

### Thoracolumbar Fascia Gain

As the abdominal wall is pulled in by the contraction of the transversus abdominis, the internal oblique acts synergistically to pull upon the thoracolumbar fascia (which wraps around the spine, connecting the deep trunk muscles to it). This in turn exerts a force on the lumbar spine which helps support and stabilize it—this force is called *thoracolumbar fascia gain*. More specifically, the increased tension of the thoracolumbar fascia compresses the erector spinae and multifidus muscles, encouraging them to contract and resist the forces that are trying to flex the spine. The classic analogy is that of the guide ropes of a tent acting together to support the main structure of the tent.

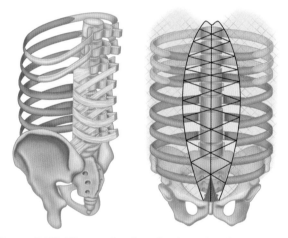

*Figure 2.36: Thoracolumbar fascia gain: the increased tension of the thoracolumbar fascia compresses the erector spinae and multifidus muscles encouraging them to contract and resist the forces that are trying to flex the spine.*

Research demonstrates that in addition to the above, the paraspinal muscles—interspinales and intertransversarii (see p. 130–133)—assist core stability, insofar as they provide an individual stabilizing effect on their adjacent vertebrae, acting in a similar way to ligaments.

It is not just the recruitment of these deep trunk muscles that is significant, but also how and when they are recruited. Hodges and Richardson (1997), two key researchers in core stability theory, showed that co-contraction of the transversus abdominis and multifidus muscles occurs prior to any movement of the limbs. This suggests that these muscles anticipate dynamic forces which may act on the lumbar spine and stabilize the area before any movement takes place elsewhere.

**Intra-abdominal Pressure (IAP)**
Pressure in the abdominal cavity is increased as a result of the abdominal wall being pulled inward by the transversus abdominis, along with a co-contraction of the pelvic floor, internal oblique, and low back muscles. This in turn exerts a tensile force on the rectus sheath, which encloses the rectus abdominis muscle. Because the rectus sheath attaches to the internal oblique and transversus abdominis muscles, it effectively surrounds the abdomen. The tension of the rectus sheath therefore increases the pressure within the abdomen like a pressurized balloon, which further facilitates the stability of the core. In practice we clearly experience this when we hold our breath during a significant lifting or throwing action, during which time we can feel ourselves contracting the diaphragm and pelvic-floor muscles.

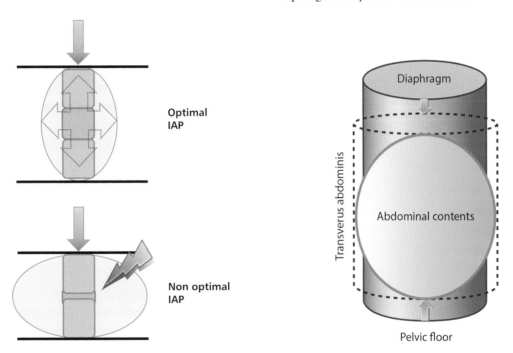

*Figure 2.37. Generation of the optimum levels of pressure inside the thoracic and abdominal cavities.*

# Biotensegrity—Biomechanics for the Twenty-first Century

John Sharkey, M.Sc. and Stephen M. Levin, M.D.

In this new edition of *The Concise Book of Muscles*, Dr. Levin and John Sharkey update the understanding of muscles in line with recent research in the field. We now know that an "origin" and "insertion" for each muscle is a simplification of the interconnectedness of muscles, ligaments, fascia, and bones. Muscles connect as a continuum to other muscles, fascial sheaths, and vascular sheaths, so their effects are dispersed and fan out over large areas and may have multiple actions (Huijing and Baan 2001). We recommend referring to "bony attachments" rather than "origin" and "insertion."

We also now know that muscles are not biphasic (i.e. either contracting or not contracting), but, instead, always have an element of tension or muscle tone (Masi and Hannon 2008). This means that the old notion of "agonist" and "antagonist" muscles should be replaced by an understanding of co-contraction of muscles and how they work in synergy. All muscles should now be considered "synergists."

Muscles that cross two or more joints become part of a closed kinematic chain, so that a muscle action remote from the target site may have significant effects at the target. An example of this is in fish jaws, where a linkage of bones may have no muscles near where the action takes place, and the more proximal muscles control the force and speed. The same can be said for the fingers and toes. There are muscles in the hand and foot, but none are strong enough alone to support your weight when you stand on your toes or hang by your fingers. It is the more central muscles that give strength to the peripheral joints through the closed kinematic chains.

This new understanding of how muscles function is consistent with the biotensegrity model that has been proposed by Levin (2002). Biotensegrity is replacing the 400-year-old compression and lever model. Biological structures are low-energy-consuming, open systems, constructed from soft, viscoelastic materials that behave nonlinearly. In this assemblage, the bones become the compressive hubs enmeshed in the complex, continuous tension network of connective tissue, fascia, ligaments, and muscles. The continuous tension network of the myofascial structure is consistent with what is now known about the soft tissues that are incorporated into the locomotor system, where tension and compression components—the soft tissue and bones—function as an integrated whole.

A new paradigm is not always easy to grasp or accept, especially when what we are being asked to comprehend is, at first glance, counterintuitive to what we see. Since the last edition of this book in 2008, researchers have provided a more accurate picture of human movement. Studies and articles from accepted experts have called into question the current paradigms of Newtonian-based biomechanics (Sharkey 2014). New research and new hypotheses help to explain how humans (and all living things) move. For example, Bartelink (1957), in an effort to provide a mathematical model to explain human movement, introduced the concept of intra-abdominal pressure. This new concept was a response to the acknowledgement that, as Levin states: "If the present paradigms of Newtonian based biomechanics hold true, then the calculated forces needed for a grandfather to lift his three-year-old grandchild would crush his spine, catching a fish at the end of a fly rod will tear the angler limb from limb, and the little sesamoid bones in our feet will crush with each step." The traditional lever-based system for explaining force production and consequent movement was recognized by many leading authorities as being flawed.

The suggestion that intra-abdominal pressure could explain how humans produce enough force to lift a heavy weight was not supported by Gracovetsky (1988). According to Gracovetsky an individual's intra-abdominal pressure would have to increase to 20 times their blood pressure (enough for them to explode) in order to allow a 250kg lift. Gracovetsky further states that the maximum load the erector spinae musculature could possibly support is no more than approximately 50kg. One hypothesis provided by Gracovetsky states that, as trunk flexion reaches the point where the erector spinae "shut down," the thoracolumbar fascia supports the load. On returning to a standing position, as the trunk nears the erect posture, the erector spinae musculature takes up the load again as the thoracolumbar fascia slackens. Contraction of the transversus abdominis draws the lateral raphe close to the tips of the spinous processes, maintaining lumbar lordosis. Such a hypothesis is a plausible alternative to the concept of intra-abdominal pressure, provided the thoracolumbar fascia can produce sufficient tensile strength to initiate such a movement. This would require forces to be translated through, or generated within, the thoracolumbar fascia.

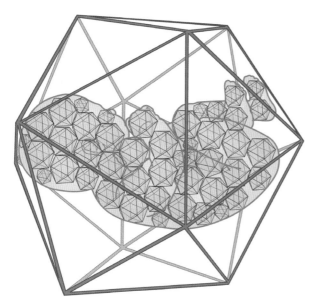

*Figure 2.38. Baby icosahedron: the image depicts multiple icosahedra within a larger icosahedron reflecting the fractal nature of biotensegrity, the living architecture or structural scaffolding of life.*

The sensitive and descriptive language of Tom Myers (2001) in *Anatomy Trains* helped the reader to appreciate the limitations and restrictions that arise from taking the approach of analyzing the body in terms of its parts, rather than viewing the human form as a unitary whole. The human body does not have bolts connecting the humerus with the ulna or linking the femur to the tibia. Cars, buildings, and non-biological structures need bolts, screws, and levers. Rather than being constructed of individual parts in a factory, the human body arose from the coming together of the largest human cell, the female egg, and the smallest human cell, the sperm. Two became four, four became eight, eight became sixteen, and so on. Each cell is a copy of its neighbor until space becomes an issue. At this point in embryonic development, cells begin to differentiate. Cells eventually specialize, becoming nerves, digestive tissue, contractile tissue, eyeballs, liver, spleen, and other "parts" that not only make up the whole but are also an integral part of the whole.

Biotensegrity counters the notion that the skeleton provides a frame for the soft tissues to hang upon, proposing instead an integrated, pre-tensioned (and self-stressed) continuous myofascial network with floating discontinuous compression struts (the skeleton) contained within it.

Architect Buckminister Fuller (Sharkey 2014) is credited with coining the term *tensegrity*, which is a combination of two words—"tension" and "integrity." Artist Kenneth Snelson, a student of Fuller, built the first floating compression structure of tensegrity in 1949. The term *biotensegrity* was coined by Dr. Levin (2002) to represent a model explaining the mechanical scaffolding that is the anatomy of all living things. The book *Biotensegrity: The Structural Basis of Life* by Scarr (2013) is highly recommended reading.

# 3 Muscles of the Scalp and Face

**Epicranius** comprises two muscle bellies (right and left) positioned opposite each other, from the front to the back of the skull, with a flat tendinous tissue linking the two. The anterior belly (or gaster) is the **frontalis**, while the posterior portion is the **occipitalis**; they are also known as the *occipitofrontalis* muscle. Epicranius, or occipitofrontalis, plays an important role in facial expression, such as raising the eyebrows.

The muscles of the external part of the ear (the auricula) include anterior, superior, and posterior portions, the largest of which is the superior auricular muscle. Similarly to the epicranius, the **auricularis anterior**, **superior**, and **posterior** are continuous with the galea aponeurosis, a flat tendon covering the skull, and they attach to the cartilage of the external ear. These muscles assist in moving the scalp and the ear.

The muscles of the eyelids include orbicularis oculi, levator palpebrae superioris, and corrugator supercilii. **Orbicularis oculi** is found circling the eye; there are three portions—namely the orbital, palpebral, and lacrimal—and these are involved in the blinking or forced closure of the eye. **Levator palpebrae superioris** is located in the orbit and, while it has a direct fascial attachment to the lacrimal gland, its main function is to assist in elevating the eyelid. **Corrugator supercilii**, associated with frowning and wrinkling of the eyebrows, is a small muscle located in the superciliary arch, drawing the eyebrows together inferiorly and medially.
The muscles of the nose include procerus, nasalis, and depressor septi nasi. **Procerus** attaches to a membrane that covers the roof of the nose and forms a bridge from the nose and the forehead, pulling the middle of the eyebrow down. It also helps the actions of the frontal bone. **Nasalis** is situated on the lateral aspect of the nose and compresses (compressor naris) and dilates (dilator nasalis) the nasal cartilages. **Depressor septi nasi**, as its name implies, draws the ala of the nose inferiorly.

There are numerous muscles of the mouth. **Orbicularis oris** is the one that encircles the mouth and lips, and is vital for facial expression and facilitating forced exhalation. **Levator labii superioris** is another muscle of facial expression, lifting the upper lip. **Levator anguli oris** has direct fascial connections with the associated zygomaticus, depressor anguli oris (triangularis), and orbicularis oris, and is therefore an important muscle of facial expression. Both **Zygomaticus major** and **minor** are facial muscles that assist in articulation of the mouth, nose, and cheeks. **Depressor labii inferioris** helps to depress the lower lip; the muscle merges with the platysma at its origin on the mandible and inserts onto the skin of the lower lip. **Depressor anguli oris** arises from the mandible of the lower jaw and inserts into the fascia of the orbicularis oris at the angle of the mouth. **Mentalis** is so named because it attaches from the mental prominence (the chin), heading upward and laterally, to attach to the soft tissue just under the lower lip; it is an important muscle of facial expression, used for expressing doubt. **Risorius** is another muscle of facial expression, taking its origin from the overlying fascia of the parotid gland; similarly to depressor anguli oris, it attaches into the modiolus and skin of the angle of the mouth. **Platysma** is a muscle of the integumentary system (horses use it to shake off irritating insects). This muscle helps to draw the mouth downward, attaching as it does to the subcutaneous fascia of the chin and jaw, down to the superficial fascia of the neck and the upper quarter of the chest (and sometimes the shoulder). **Buccinator** is an important muscle of mastication and in facial expressions, such as smiling; a newborn baby uses this muscle for sucking.

The muscles of mastication include masseter, temporalis, and the pterygoids. **Masseter** is a major muscle used for chewing and for elevating and protracting the mandible. Arising from the zygomatic process of the maxilla and two-thirds of the zygomatic arch, this muscle inserts at the angle of the mandible and the outer surface of the ramus and coronoid process of the mandible. Synergistic with masseter, **temporalis** arises from the temporal fascia covering the zygomatic, frontal, parietal, sphenoid, and temporal bones, and inserts on the apex (medial/lateral) of the coronoid process of the mandible and anterior border of the ramus of the mandible. A short spastic temporalis leads to teeth clenching, causing damage to the sensitive proprioceptive covering of the teeth. The **pterygoids** (medial and lateral) act to elevate the mandible and close the jaw, while the lateral pterygoid moves the jaw from side to side. These muscles arise from the medial side of the lateral pterygoid plate and the superficial head of the maxillary tuberosity; their insertion is onto the medial (fovea) angle of the mandible.

For the most part, the muscles mentioned above are innervated by the facial nerve or one of its anterior and posterior contributions, such as the trigeminal **V** nerve.

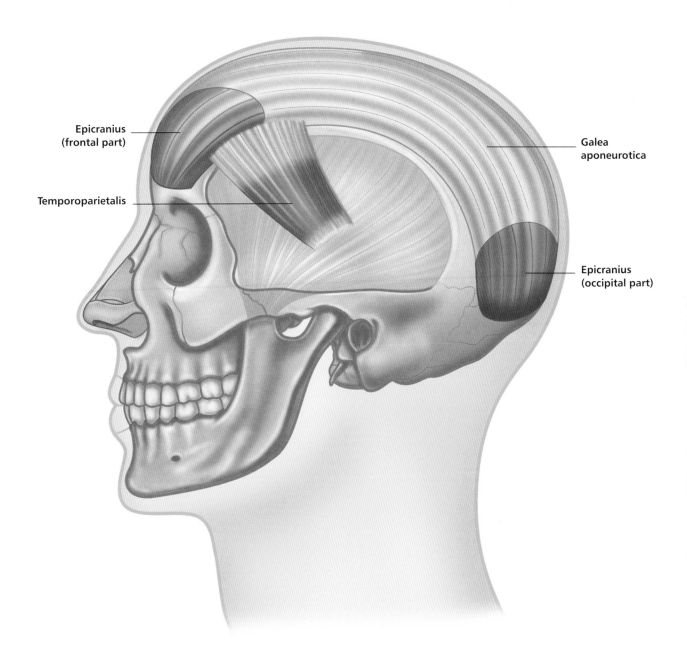

Epicranius
(frontal part)

Temporoparietalis

Galea
aponeurotica

Epicranius
(occipital part)

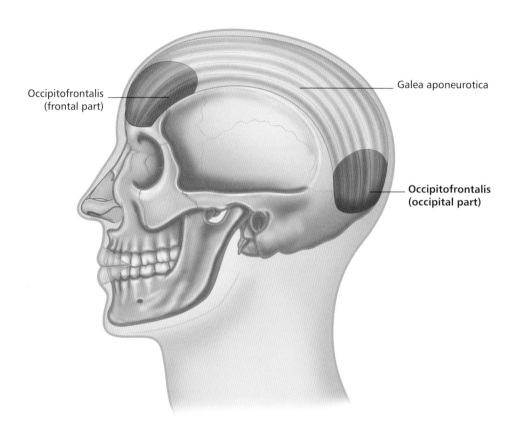

Occipitofrontalis
(frontal part)

Galea aponeurotica

Occipitofrontalis
(occipital part)

**Latin**, *occiput*, back of the head.

The epicranius (occipitofrontalis) is effectively two muscles (occipitalis and frontalis) united by an aponeurosis called the *galea aponeurotica*, so named because it forms what resembles a helmet (**Latin** *galea*).

**Origin**
Lateral two-thirds of superior nuchal line of occipital bone. Mastoid process of temporal bone.

**Insertion**
Galea aponeurotica (a sheet-like tendon leading to frontal belly).

**Action**
Pulls scalp backward. Assists frontal belly in raising eyebrows and wrinkling forehead.

**Nerve**
Facial **VII** nerve (posterior auricular branch).

**Basic functional movement**
Facilitates facial expressions.

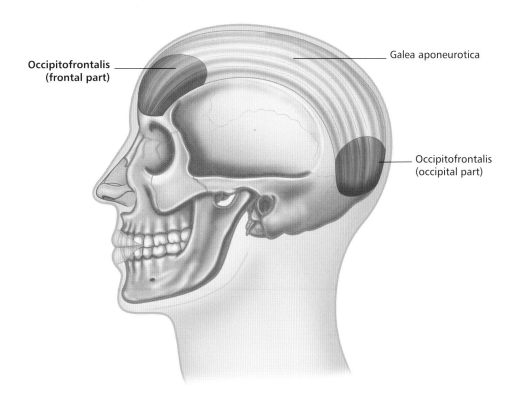

Occipitofrontalis
(frontal part)

Galea aponeurotica

Occipitofrontalis
(occipital part)

**Latin**, *frons*, forehead, front of the head.

The epicranius (occipitofrontalis) is effectively two muscles (occipitalis and frontalis), united by an aponeurosis called the *galea aponeurotica*, so named because it forms what resembles a helmet (**Latin**, *galea*).

**Origin**
Galea aponeurotica (a sheet-like tendon leading to occipital belly).

**Insertion**
Fascia and skin above eyes and nose.

**Action**
Pulls scalp forward. Raises eyebrows and wrinkles skin of forehead horizontally.

**Nerve**
Facial **VII** nerve (temporal branches).

**Basic functional movement**
Facilitates facial expressions.

# TEMPOROPARIETALIS

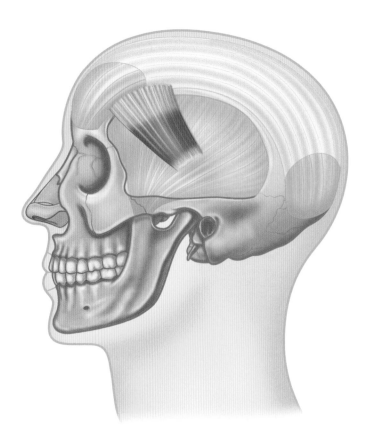

**Latin**, *tempus*, temple; *parietalis*, relating to the walls of a cavity.

**Origin**
Fascia above ear.

**Insertion**
Lateral border of galea aponeurotica.

**Action**
Tightens scalp. Raises ears.

**Nerve**
Facial **VII** nerve (temporal branch).

# Muscles of the Ear

The auricularis anterior, superior, and posterior are also referred to as the *extrinsic muscles of the auricle*. They are generally non-functional in humans unless trained.

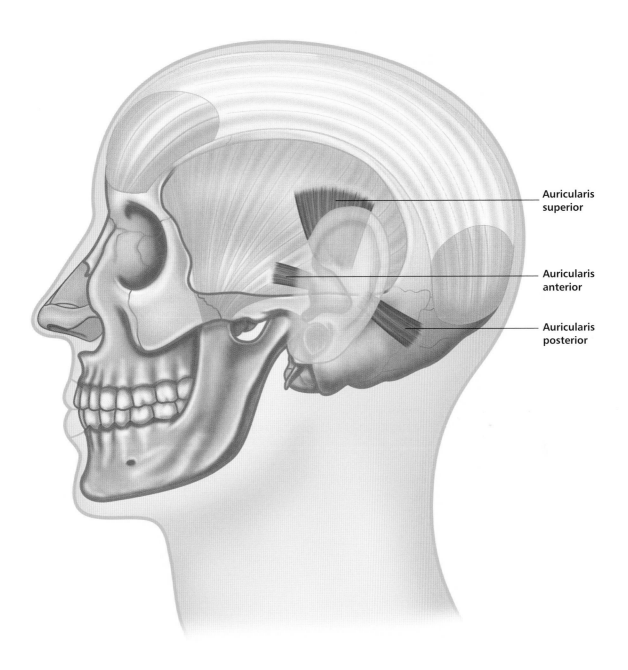

Auricularis superior

Auricularis anterior

Auricularis posterior

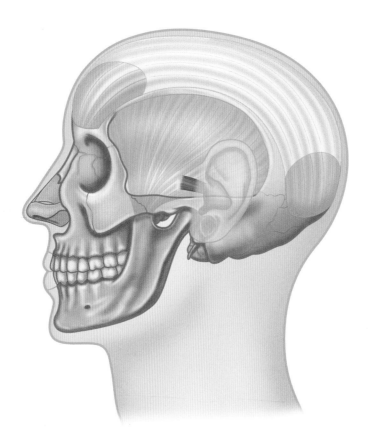

**Latin**, *auricularis*, relating to the ear; *anterior*, at the front.

**Origin**
Fascia in temporal region anterior to ear.

**Insertion**
Anterior to helix of ear.

**Action**
Draws ear forward. Moves scalp.

**Nerve**
Facial **VII** nerve (temporal branch).

# AURICULARIS SUPERIOR

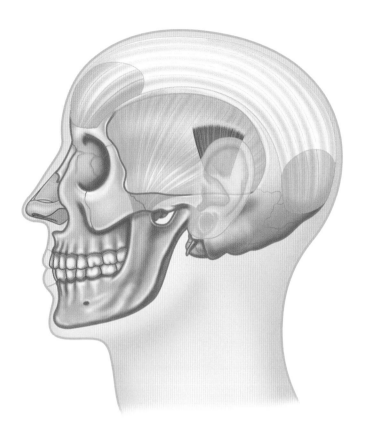

**Latin**, *auricularis*, relating to the ear; *superior*, upper.

**Origin**
Fascia in temporal region above ear.

**Insertion**
Superior part of ear.

**Action**
Draws ear forward. Moves scalp.

**Nerve**
Facial **VII** nerve (temporal branch).

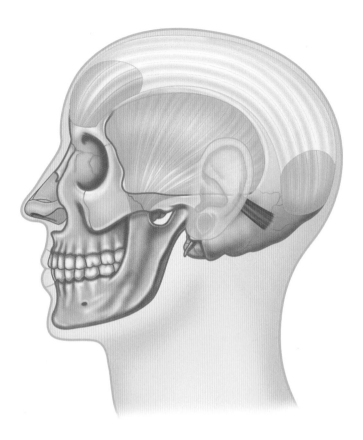

**Latin**, *auricularis*, relating to the ear; *posterior*, at the back.

**Origin**
Temporal bone, near mastoid process.

**Insertion**
Posterior part of ear.

**Action**
Pulls ear upward.

**Nerve**
Facial **VII** nerve (posterior auricular branch).

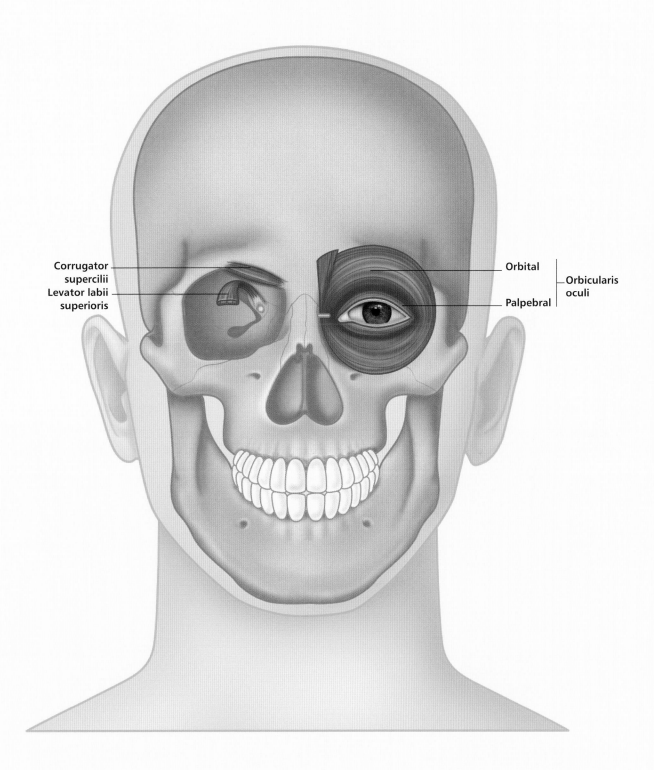

Corrugator
supercilii

Levator labii
superioris

Orbital

Palpebral

Orbicularis
oculi

# ORBICULARIS OCULI

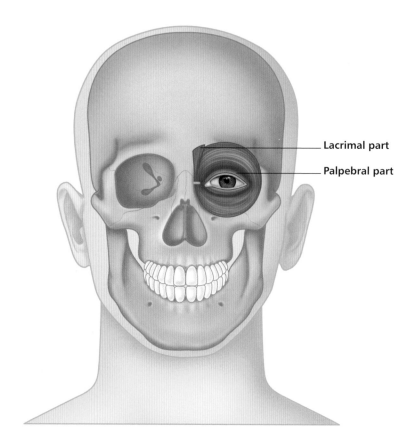

Lacrimal part

Palpebral part

**Latin**, *orbiculus*, small circular disc; *oculus*, eye.

This complex and extremely important muscle consists of three parts—lacrimal, orbital, and palpebral; together they form an important protective mechanism surrounding the eye.

**LACRIMAL PART**
(behind medial palpebral ligament and lacrimal sac)

**Latin**, *lacrima*, tear.

**Origin**
Lacrimal bone.

**Insertion**
Lateral palpebral raphe.

**Action**
Dilates lacrimal sac and brings lacrimal canals onto surface of eye.

**Nerve**
Facial **VII** nerve (temporal and zygomatic branches).

**ORBITAL PART**
(circling the eye)

**Origin**
Frontal bone. Medial wall of orbit (on maxilla).

**Insertion**
Circular path around orbit, returning to origin.

**Action**
Strongly closes eyelids (firmly "screws up" eye).

**Nerve**
Facial **VII** nerve (temporal and zygomatic branches).

**PALPEBRAL PART**
(in eyelids)

**Latin**, *palpebra*, eyelid.

**Origin**
Medial palpebral ligament.

**Insertion**
Lateral palpebral ligament into zygomatic bone.

**Action**
Gently closes eyelids (and comes into action involuntarily, as in blinking).

**Nerve**
Facial **VII** nerve (temporal and zygomatic branches).

# LEVATOR PALPEBRAE SUPERIORIS

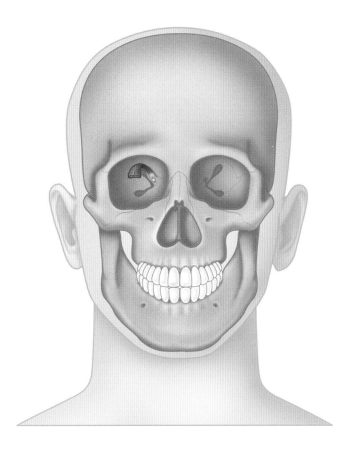

**Latin**, *levare*, to lift; *palpebrae*, of the eyelid; *superioris*, of the upper.

This muscle is unusual in that it contains both somatic and visceral muscle fibers. It is the antagonist of the palpebral part of orbicularis oculi; therefore, paralysis of levator palpebrae superioris results in the upper eyelid drooping down over the eyeball.

**Origin**
Root of orbit (lesser wing of sphenoid bone).

**Insertion**
Skin of upper eyelid.

**Action**
Raises upper eyelid.

**Nerve**
Oculomotor **III** nerve.

**Basic functional movement**
Waking up.

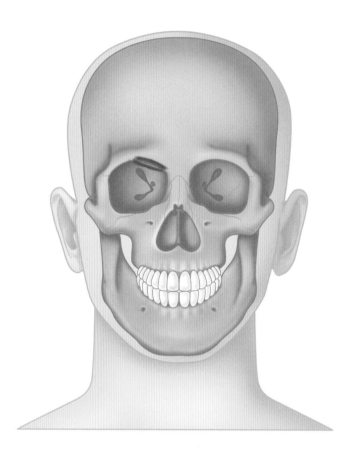

**Latin**, *corrugare*, to wrinkle up;
*supercilii*, of the eyebrow.

**Origin**
Medial end of superciliary arch of
frontal bone.

**Insertion**
Deep surface of skin under medial
half of eyebrows.

**Action**
Draws eyebrows medially and
downward, thus producing vertical
wrinkles, as in frowning.

**Nerve**
Facial **VII** nerve (temporal
branch).

**Basic functional movement**
Facilitates facial expression.

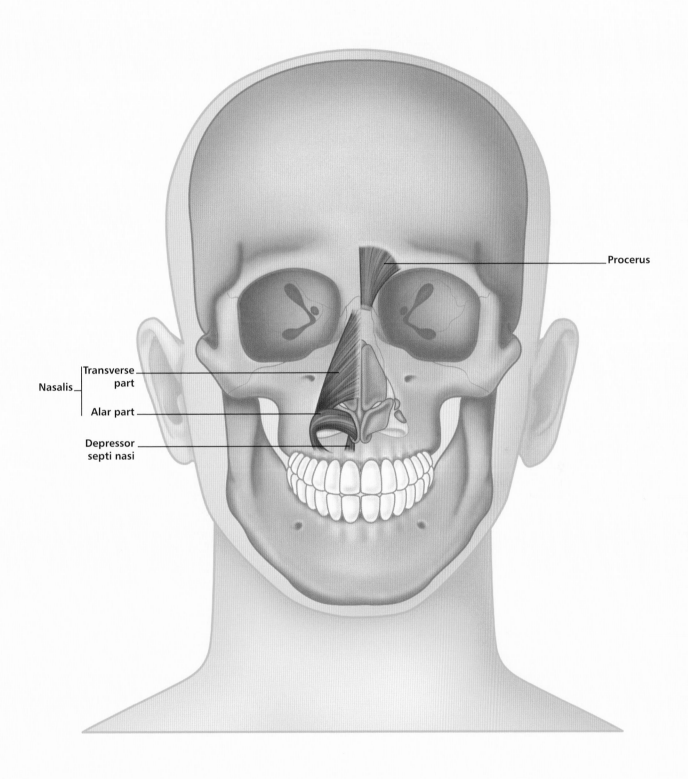

Procerus

Nasalis — Transverse part

Alar part

Depressor septi nasi

# PROCERUS

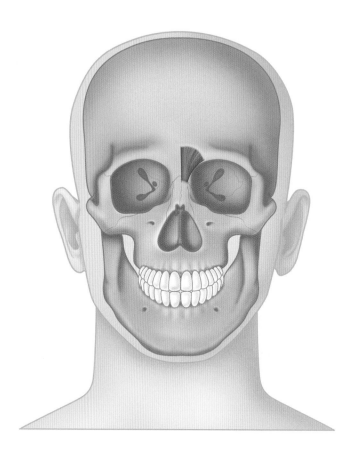

**Latin**, *procerus*, long.

**Origin**
Fascia over nasal bone. Lateral nasal cartilage.

**Insertion**
Skin between eyebrows.

**Action**
Wrinkles nose. Pulls medial portion of eyebrows downward.

**Nerve**
Facial **VII** nerve (temporal branches).

**Basic functional movement**
Example: enables strong "sniffing" and sneezing.

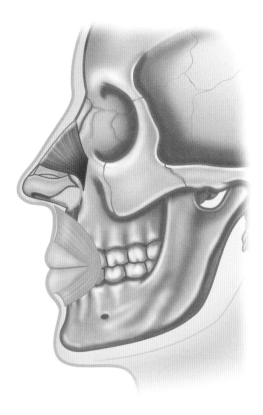

**Latin**, *nasus*, nose.

**Origin**
Middle of maxilla (above incisor and canine teeth). Greater alar cartilage. Skin on nose.

**Insertion**
Joins muscle of opposite side across bridge of nose. Skin at tip of nose.

**Action**
Maintains opening of external nares during forceful inhalation (i.e. flares nostrils).

**Nerve**
Facial **VII** nerve (buccal branches).

**Basic functional movement**
Example: breathing in strongly through nose.

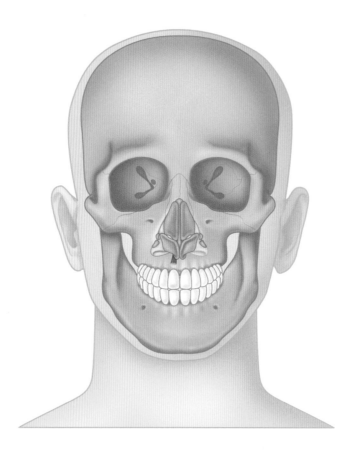

**Latin**, *deprimere*, to press down; *septi*, of the dividing wall; *nasi*, of the nose.

**Origin**
Incisive fossa of maxilla (above incisor teeth).

**Insertion**
Nasal septum and ala.

**Action**
Constricts nares.

**Nerve**
Facial **VII** nerve (buccal branches).

**Basic functional movement**
Example: twitching the nose.

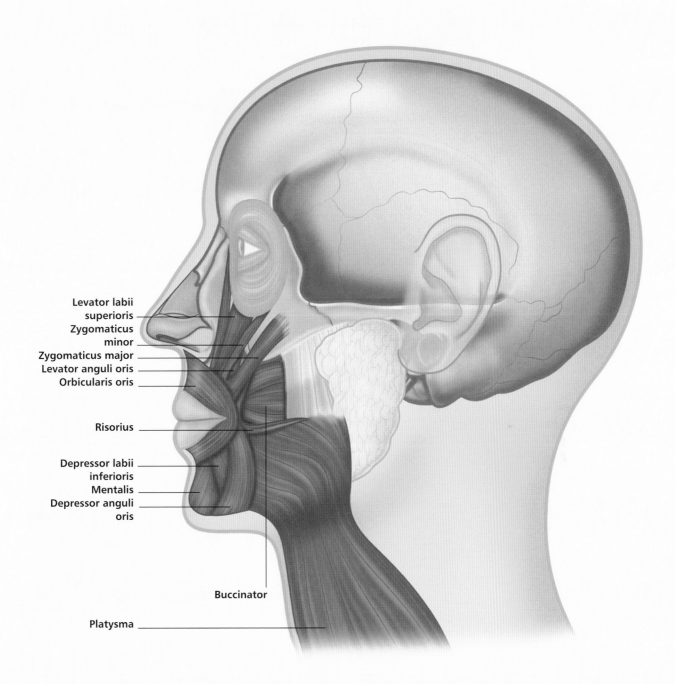

Levator labii
superioris

Zygomaticus
minor

Zygomaticus major

Levator anguli oris

Orbicularis oris

Risorius

Depressor labii
inferioris

Mentalis

Depressor anguli
oris

Buccinator

Platysma

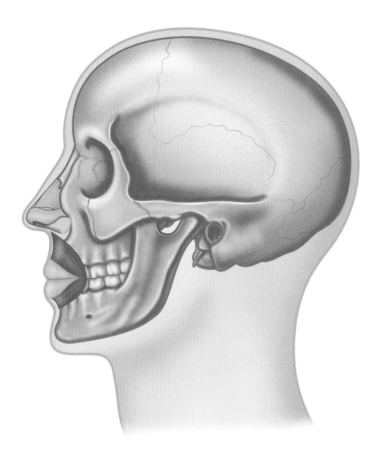

**Latin**, *orbiculus*, small circular disc; *oris*, of the mouth.

This is a composite sphincter muscle that encircles the mouth; it receives fasciculi from many other muscles.

### Origin
Muscle fibers surrounding opening of mouth, attached to skin, muscle, and fascia of lips and surrounding area.

### Insertion
Skin and fascia at corner of mouth.

### Action
Closes lips. Compresses lips against teeth. Protrudes (purses) lips. Shapes lips during speech.

### Nerve
Facial **VII** nerve (buccal and mandibular branches).

### Basic functional movement
Facial expressions involving the lips.

# LEVATOR LABII SUPERIORIS

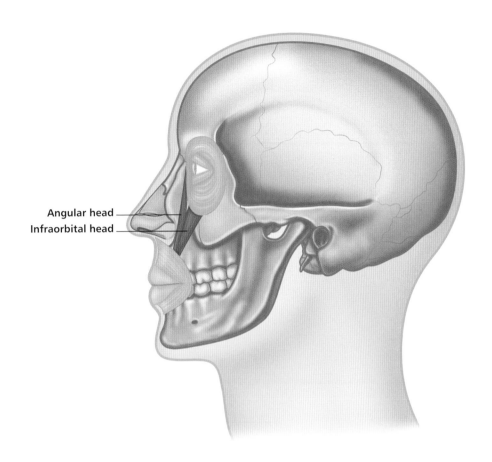

Angular head
Infraorbital head

Latin, *levare*, to lift; *labii*, of the lip; *superioris*, of the upper.

## ANGULAR HEAD

**Origin**
Zygomatic bone and frontal process of maxilla.

**Insertion**
Greater alar cartilage, upper lip, and skin of nose.

**Action**
Raises upper lip. Dilates nares. Forms nasolabial furrow.

**Nerve**
Facial **VII** nerve (buccal branches).

**Basic functional movement**
Facilitates facial expression and kissing.

## INFRAORBITAL HEAD

**Origin**
Lower border of orbit.

**Insertion**
Muscles of upper lip.

**Action**
Raises upper lip.

**Nerve**
Facial **VII** nerve (buccal branches).

**Basic functional movement**
Facilitates facial expression and kissing.

# LEVATOR ANGULI ORIS

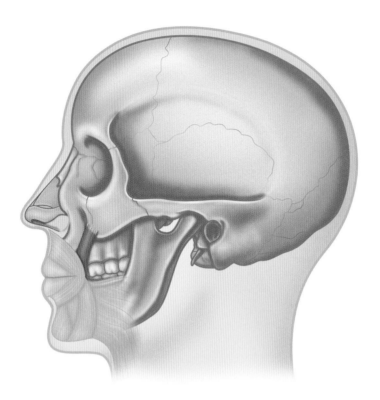

**Latin**, *levare*, to lift; *anguli*, of the corner; *oris*, of the mouth.

**Origin**
Canine fossa of maxilla.

**Insertion**
Corner of mouth.

**Action**
Elevates angle (corner) of mouth.

**Nerve**
Facial **VII** nerve (buccal branches).

**Basic functional movement**
Helps produce a smiling expression.

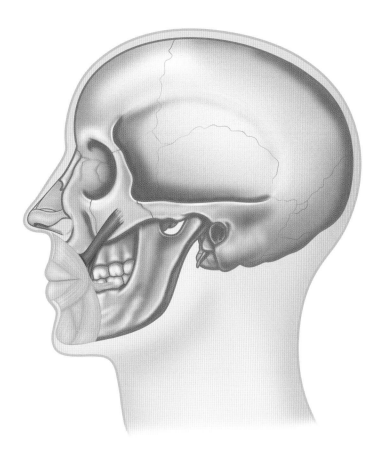

**Greek**, *zygoma*, bar, bolt. **Latin**, *major*, larger.

**Origin**
Upper lateral surface of zygomatic bone.

**Insertion**
Skin at corner of mouth. Orbicularis oris.

**Action**
Pulls corner of mouth up and back, as in smiling.

**Nerve**
Facial **VII** nerve (zygomatic and buccal branches).

**Basic functional movement**
Smiling.

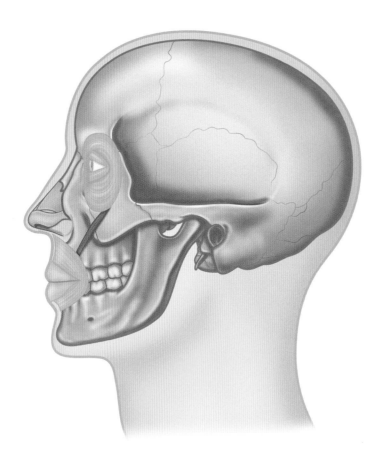

**Greek**, *zygoma*, bar, bolt. **Latin**, *minor*, smaller.

**Origin**
Lower surface of zygomatic bone.

**Insertion**
Lateral part of upper lip, lateral to levator labii superioris.

**Action**
Elevates upper lip. Forms nasolabial furrow.

**Nerve**
Facial **VII** nerve (buccal branches).

**Basic functional movement**
Facilitates facial expression.

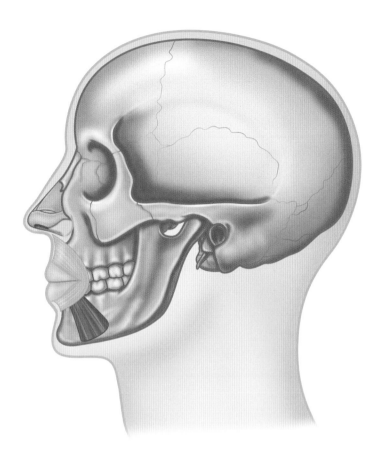

**Latin**, *deprimere*, to press down; *labii*, of the lip; *inferioris*, of the lower.

**Origin**
Anterior surface of mandible, between mental foramen and symphysis.

**Insertion**
Skin of lower lip.

**Action**
Pulls lower lip downward and slightly laterally.

**Nerve**
Facial **VII** nerve (marginal mandibular branch).

**Basic functional movement**
Facilitates facial expression.

# DEPRESSOR ANGULI ORIS

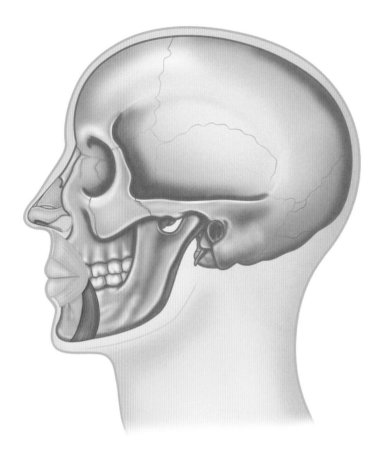

**Latin**, *deprimere*, to press down; *anguli*, of the corner; *oris*, of the mouth.

Muscle fibers are continuous with platysma.

**Origin**
Oblique line of mandible.

**Insertion**
Corner of mouth.

**Action**
Pulls corner of mouth downward, as in sadness or frowning.

**Nerve**
Facial **VII** nerve (marginal mandibular and buccal branches).

# MENTALIS

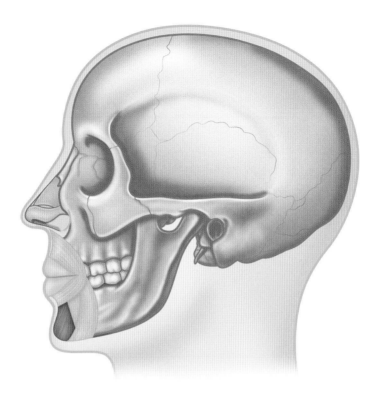

**Latin**, *mentum*, chin.

This is the only muscle of the lips that normally has no connection with the orbicularis oris.

**Origin**
Incisive fossa of anterior surface of mandible.

**Insertion**
Skin of chin.

**Action**
Protrudes lower lip and pulls up (wrinkles) skin of chin, as in pouting.

**Nerve**
Facial **VII** nerve (marginal mandibular branch).

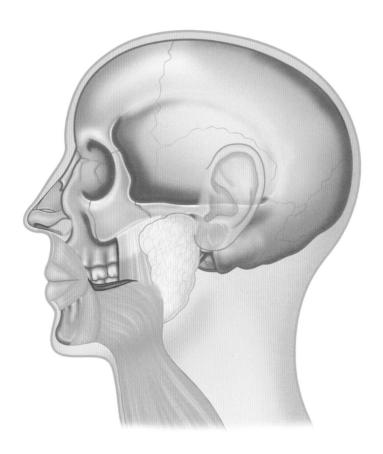

**Latin**, *risus*, laughter.

This thin muscle is often completely fused with platysma.

**Origin**
Fascia over masseter and parotid (salivary) gland (i.e. fascia of lateral cheek).

**Insertion**
Skin at angle of mouth.

**Action**
Draws angle of mouth laterally, as in tenseness or grinning.

**Nerve**
Facial **VII** nerve (buccal branches).

# PLATYSMA

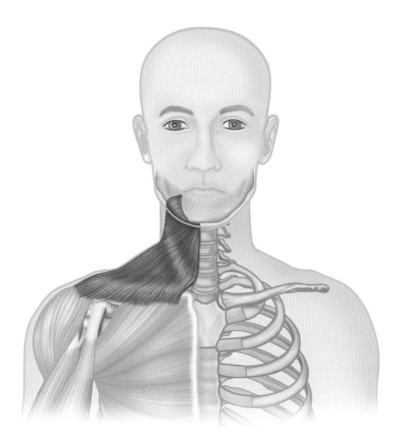

**Greek**, *platys*, broad, flat.

This muscle may be seen to stand out in a runner finishing a hard race.

**Origin**
Subcutaneous fascia of upper quarter of chest (i.e. fascia overlying pectoralis major and deltoid muscles).

**Insertion**
Subcutaneous fascia and muscles of chin and jaw. Inferior border of mandible.

**Action**
Pulls lower lip from corner of mouth downward and laterally. Draws skin of chest upward.

**Nerve**
Facial **VII** nerve (cervical branch).

**Basic functional movement**
Example: produces expression of being startled or of sudden fright.

# BUCCINATOR

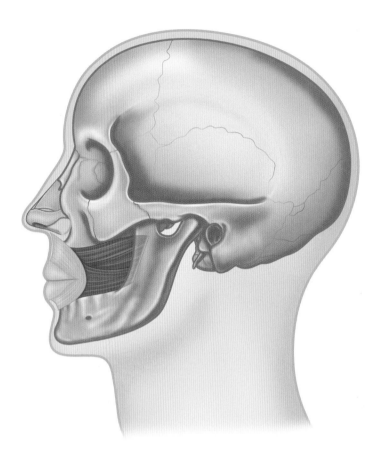

**Latin**, *bucca*, cheek.

This muscle forms the substance of the cheek.

### Origin
Alveolar processes of maxilla and mandible over molars and along pterygomandibular raphe (fibrous band extending from pterygoid hamulus to mandible).

### Insertion
Orbicularis oris (muscles of lips).

### Action
Compresses cheek, as in blowing air out of mouth. Caves cheeks in, producing the action of sucking.

### Nerve
Facial **VII** nerve (buccal branches).

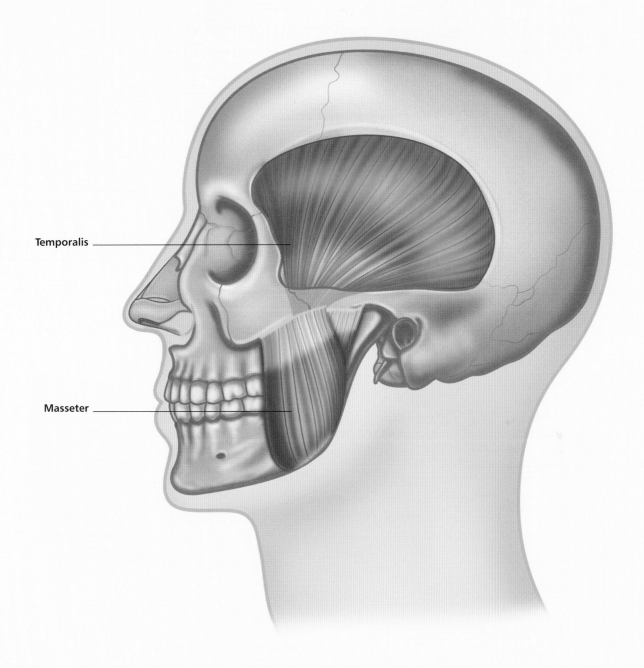

Temporalis

Masseter

# MASSETER

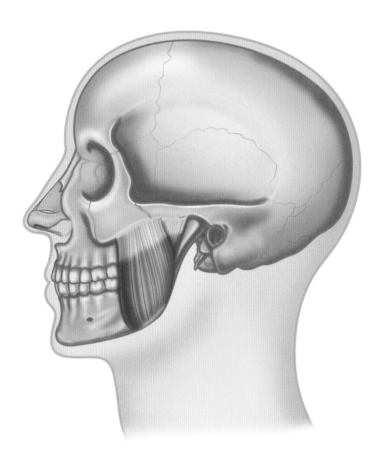

**Greek**, *maseter*, chewer.

The masseter is the most superficial muscle of mastication, easily felt when the jaw is clenched.

**Origin**
Zygomatic process of maxilla. Medial and inferior surfaces of zygomatic arch.

**Insertion**
Angle of ramus of mandible. Coronoid process of mandible.

**Action**
Closes jaw. Clenches teeth. Assists in side-to-side movement of mandible.

**Nerve**
Trigeminal **V** nerve (mandibular division).

**Basic functional movement**
Chewing food.

# TEMPORALIS

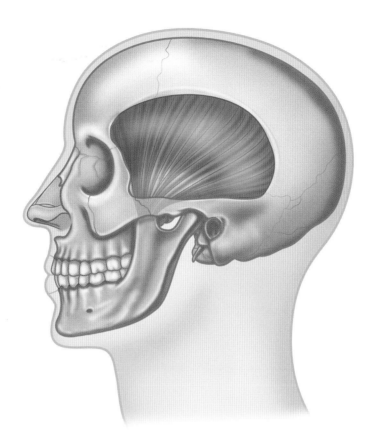

**Latin**, *temporalis*, relating to the side of the head.

Temporalis is a broad fan-shaped muscle and covers much of the temporal bone.

**Origin**
Temporal fossa, including parietal, temporal, and frontal bones. Temporal fascia.

**Insertion**
Coronoid process of mandible. Anterior border of ramus of mandible.

**Action**
Closes jaw. Clenches teeth. Assists in side-to-side movement of mandible.

**Nerve**
Anterior and posterior deep temporal nerves from the trigeminal **V** nerve (mandibular division).

**Basic functional movement**
Chewing food.

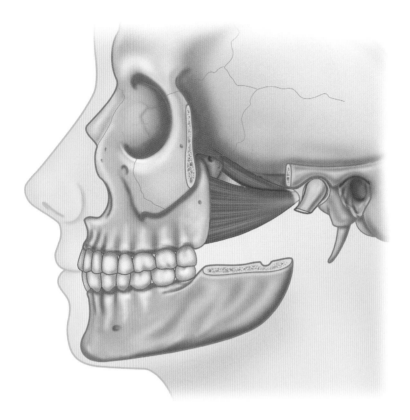

**Greek**, *pterygoeides*, wing-like.
**Latin**, *lateralis*, relating to the side.

The superior head of this muscle is sometimes called *sphenomeniscus*, because it inserts into the disc of the temporomandibular joint.

**Origin**
Superior head: lateral surface of greater wing of sphenoid.
Inferior head: lateral surface of lateral pterygoid plate of sphenoid.

**Insertion**
Superior head: capsule and articular disc of temporomandibular joint.
Inferior head: neck of mandible.

**Action**
Protrudes mandible. Opens mouth. Moves mandible from side to side (as in chewing).

**Nerve**
Trigeminal **V** nerve (mandibular division).

**Basic functional movement**
Chewing food.

# PTERYGOIDEUS MEDIALIS (Medial Pterygoid)

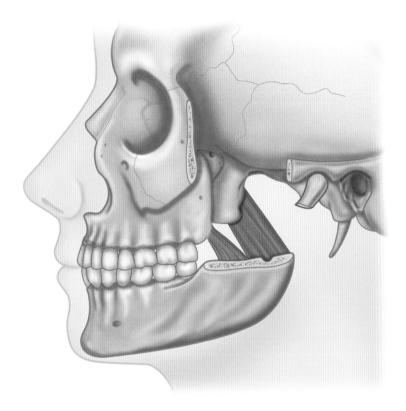

**Greek**, *pterygoeides*, wing-like. **Latin**, *medialis*, relating to the middle.

This muscle mirrors the masseter muscle in both its position and action, with the ramus of the mandible positioned between the two muscles.

**Origin**
Medial surface of lateral pterygoid plate of sphenoid bone. Pyramidal process of palatine bone. Tuberosity of maxilla.

**Insertion**
Medial surface of ramus and angle of mandible.

**Action**
Elevates and protrudes mandible, thus closing jaw and assisting in side-to-side movement of mandible, as in chewing.

**Nerve**
Trigeminal **V** nerve (mandibular division).

**Basic functional movement**
Chewing food.

# Muscles of the Neck

The **hyoids**, along with **digastricus**, attach to, and facilitate the positioning of, the hyoid bone, which is the only bone in the human body that has no approximation to any other bone. These muscles, along with the specific location of the hyoid bone, are essential for speech, providing the tongue with appropriate tensional support and moving the larynx as we speak or swallow. The muscles of the hyoid group have attachments to the mandible, the temporal bone, the manubrium, the clavicle, the costal cartilage of the first rib, and the thyroid cartilage.

The anterior vertebral muscles are a small group of muscles attached to the bodies and transverse processes of the cervical and upper thoracic regions of the vertebral column. **Longus colli** has three specific parts (superior oblique, inferior oblique, vertical), which lie on the anterior lateral aspect of both the upper cervical and thoracic vertebrae; they range from the transverse processes of the third, fourth, and fifth cervical vertebrae, with attachments to the anterior aspects of the first two cervical vertebrae and including the anterior surface of the first three thoracic vertebrae. The muscle inserts at the anterior tubercle of the atlas and anterior tubercles of the transverse processes C5, 6. **Longus capitis** originates from the anterior tubercles of the transverse processes of the third, fourth, fifth, and sixth cervical vertebrae and inserts into the inferior surface of the basilar portion of the occiput. The muscle decelerates neck extension. **Rectus capitis anterior** and **rectus capitis lateralis** decelerate the head during extension and contralateral flexion, as the anterior muscle originates from the anterior surface of the lateral mass of the atlas, while the lateralis originates from the transverse process of the atlas. These two muscles insert into the basilar (anterior) and jugular (lateralis) portions.

The lateral vertebral muscles of the neck comprise the scalenes (scalenus anterior, medius, and posterior), which run from the transverse processes of the cervical vertebrae downward to the ribs, and the sternocleidomastoideus. The **scalenes** originate from the transverse process of all the cervical vertebrae and insert on the first rib and/or suprapleural membrane. The posterior portion may attach to the first two ribs. These muscles elevate the ribs for respiration if the ribs are fixed, rotate to the side opposite of the muscle contracting, and laterally flex to the contracted side. Contraction of the muscles on both sides flexes the neck. **Sternocleidomastoideus** (SCM) originates from the manubrium of the sternum and the medial portion of the clavicle (two heads), to insert on the mastoid process of the temporal bone. This muscle is a key player in head positioning: contraction of the muscle causes rotation to the side opposite that contracting and lateral flexion to the contracted side. Bilateral contraction of the muscle flexes the cervical spine (neck).

# Hyoid Muscles

The hyoid muscles are mostly concerned with steadying or moving the hyoid bone, and therefore the tongue and larynx, which are attached to it. The **suprahyoid** muscles include: mylohyoideus, geniohyoideus, stylohyoideus, and digastric. The suprahyoid muscles are located above (superior to) the hyoid bone. The **infrahyoid** muscles include: sternohyoideus, sternothyroideus, thyrohyoideus, and omohyoideus. The infrahyoid muscles are located below (inferior to) the hyoid bone.

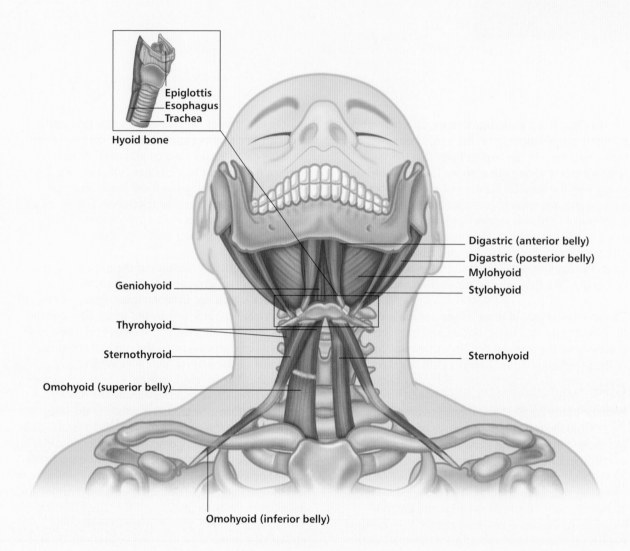

Epiglottis
Esophagus
Trachea

Hyoid bone

Digastric (anterior belly)
Digastric (posterior belly)
Mylohyoid
Stylohyoid

Geniohyoid

Thyrohyoid

Sternothyroid

Sternohyoid

Omohyoid (superior belly)

Omohyoid (inferior belly)

# MYLOHYOIDEUS

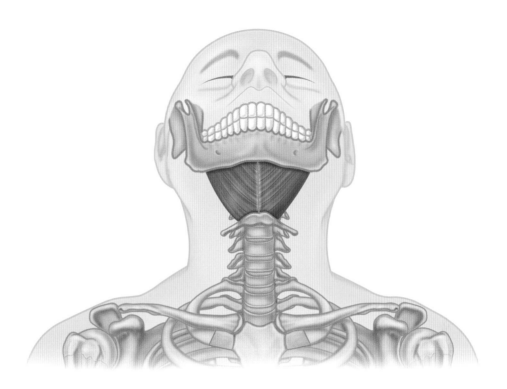

**Greek**, *mylos*, millstone, molar; *hyoeides*, shaped like the Greek letter upsilon (υ).

The mylohyoid fibers form a sling or diaphragm that supports the floor of the mouth.

**Origin**
Mylohyoid line on inner surface of mandible.

**Insertion**
Hyoid bone.

**Action**
Raises floor of mouth in swallowing. Elevates hyoid bone. Helps press tongue upward and backward against roof of mouth.

**Nerve**
Mylohyoid nerve from inferior alveolar nerve. This is a branch of trigeminal **V** nerve (mandibular division).

**Basic functional movement**
Swallowing.

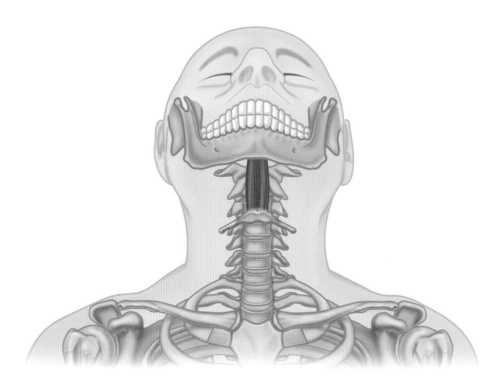

**Greek**, *geneion*, chin; *hyoeides*, shaped like the Greek letter upsilon (υ).

**Origin**
Lower part of mental spine of interior medial surface of mandible.

**Insertion**
Hyoid bone.

**Action**
Protrudes and elevates hyoid bone, widening pharynx for reception of food. Can help retract and depress mandible if hyoid bone is fixed.

**Nerve**
Fibers of cervical nerve C1, conveyed by the hypoglossal nerve **XII**.

# STYLOHYOIDEUS

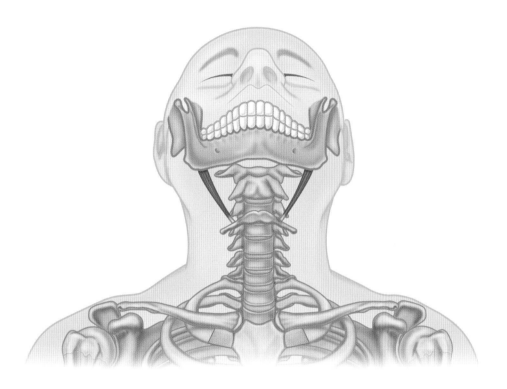

**Latin**, *stilus*, stake, pale. **Greek**, *hyoeides*, shaped like the Greek letter upsilon (υ).

**Origin**
Posterior border of styloid process of temporal bone.

**Insertion**
Hyoid bone (after splitting to enclose the intermediate tendon of digastric).

**Action**
Pulls hyoid bone upward and backward, thereby elevating tongue.

**Nerve**
Facial **VII** nerve (mandibular branches).

# DIGASTRICUS

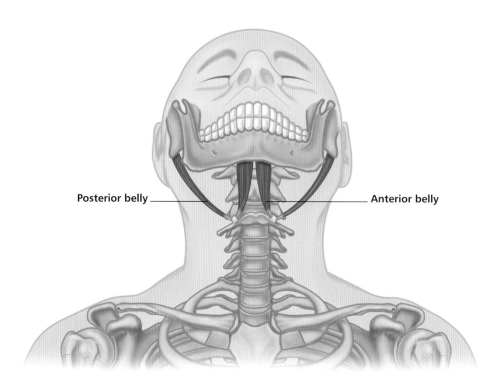

Posterior belly _____ Anterior belly

**Latin**, *digastricus*, having two (muscle) bellies.

The digastricus consists of two muscular bellies, the anterior belly and the posterior belly, conjoined by an intermediate tendon.

### Origin
Anterior belly: digastric fossa on inner side of lower border of mandible, near symphysis. Posterior belly: mastoid notch of temporal bone.

### Insertion
Body of hyoid bone via a fascial sling over an intermediate tendon.

### Action
Raises hyoid bone. Depresses and retracts mandible, as in opening mouth.

### Nerve
Anterior belly: mylohyoid nerve, from trigeminal **V** nerve (mandibular division). Posterior belly: facial **VII** nerve (digastric branch).

# STERNOHYOIDEUS

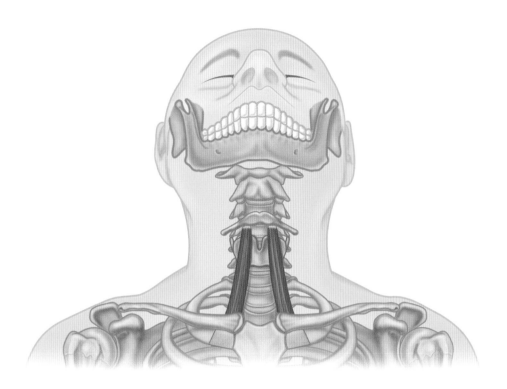

**Greek**, *sternon*, chest; *hyoeides*, shaped like the Greek letter upsilon (υ).

**Origin**
Posterior surface of manubrium of sternum. Medial end of clavicle.

**Insertion**
Lower border of hyoid bone (medial to insertion of omohyoideus).

**Action**
Depresses hyoid bone. Stabilizes hyoid bone when other muscles are acting from it.

**Nerve**
Ansa cervicalis nerve C1, **2**, **3**.

# STERNOTHYROIDEUS

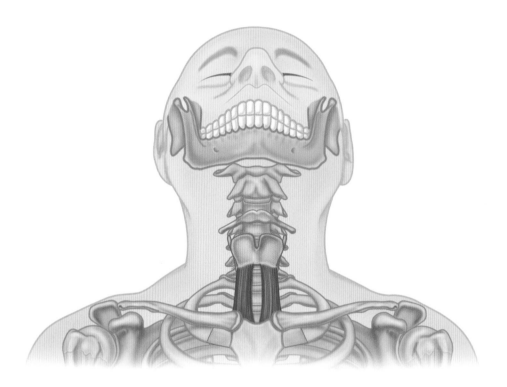

**Greek**, *sternon*, chest; *thyreos*, oblong shield.

Lies deep to sternohyoideus.

### Origin
Posterior surface of manubrium of sternum, below origin of sternohyoideus. First costal cartilage.

### Insertion
Oblique line on outer surface of thyroid cartilage.

### Action
Pulls thyroid cartilage away from hyoid bone, thus opening laryngeal orifice.

### Nerve
Ansa cervicalis nerve C**1**, **2**, **3**.

# THYROHYOIDEUS

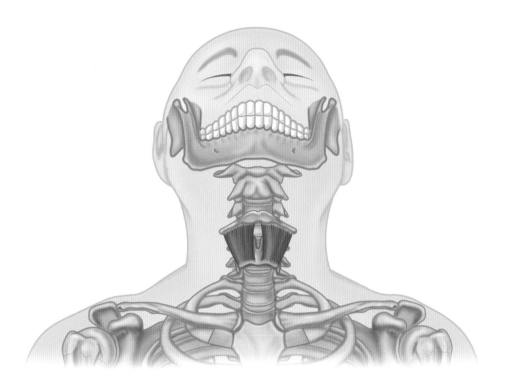

**Greek**, *thyreos*, oblong shield; *hyoeides*, shaped like the Greek letter upsilon (υ).

This is a short strap muscle.

**Origin**
Oblique line of outer surface of thyroid cartilage.

**Insertion**
Lower border of body and greater horn of hyoid bone.

**Action**
Raises thyroid and depresses hyoid bone, thus closing laryngeal orifice, preventing food from entering larynx during swallowing.

**Nerve**
Ansa cervicalis nerve C**1**, 2, via fibers from descending hypoglossal **XII** nerve.

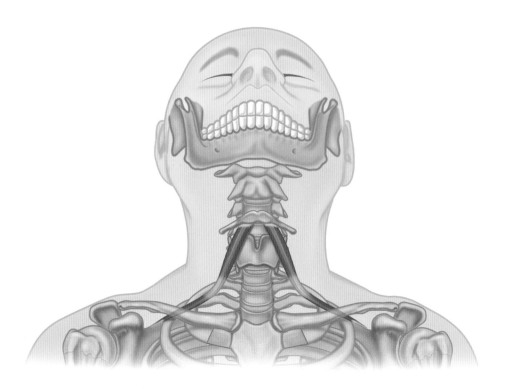

**Greek**, *omos*, shoulder; *hyoeides*, shaped like the Greek letter upsilon (υ).

## Origin
Inferior belly: upper border of scapula, medial to the scapular notch. Superior transverse ligament.
Superior belly: intermediate tendon.

## Insertion
Inferior belly: intermediate tendon.
Superior belly: lower border of hyoid bone, lateral to insertion of sternohyoideus.

Note: The intermediate tendon is tied down to the clavicle and first rib by a sling of the cervical fascia.

## Action
Depresses hyoid bone.

## Nerve
Ansa cervicalis nerve C2, 3.

# Anterior Vertebral Muscles

The anterior vertebral muscles are a small group of muscles attached to the bodies and transverse processes of the cervical and upper thoracic regions of the vertebral column.

**Strengthen**

*Isometric forward neck exercise*

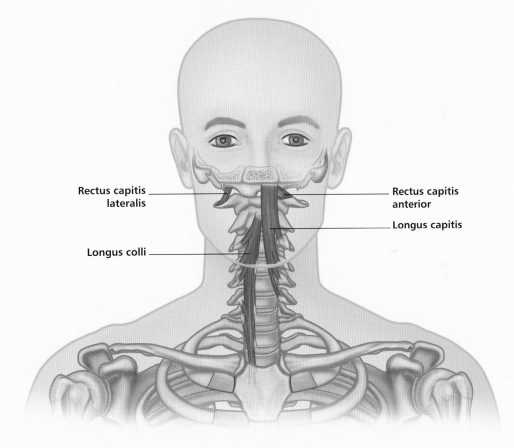

Rectus capitis lateralis

Rectus capitis anterior

Longus capitis

Longus colli

**Self-stretch**

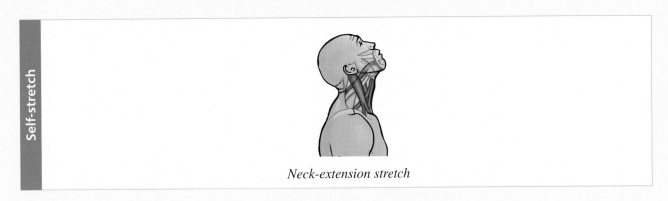

*Neck-extension stretch*

# LONGUS COLLI

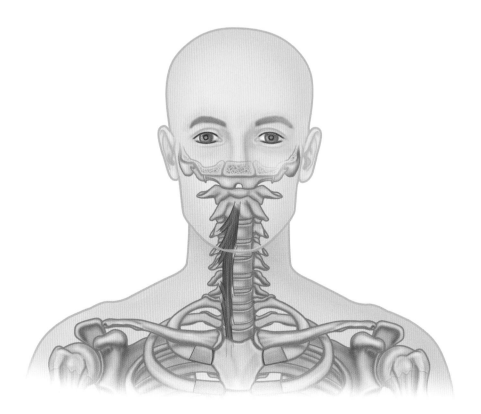

**Latin**, *longus*, long; *colli*, of the neck.

Longus colli can be divided into three parts—superior oblique, inferior oblique, and vertical—and is the largest member of the prevertebral muscles.

## SUPERIOR OBLIQUE PART
**Origin**
Transverse processes of third, fourth, and fifth cervical vertebrae (C3–5).

**Insertion**
Anterior arch of atlas.

**Action**
Flexes cervical vertebrae.

**Nerve**
Ventral rami of cervical nerves C2–7.

**Basic functional movement**
Gives smoothness and stability to flexion at the neck.

## INFERIOR OBLIQUE PART
**Origin**
Anterior surface of first two or three cervical vertebral bodies.

**Insertion**
Transverse processes of fifth and sixth cervical vertebrae (C5–6).

**Action**
Flexes cervical vertebrae.

**Nerve**
Ventral rami of cervical nerves C2–7.

**Basic functional movement**
Gives smoothness and stability to flexion at the neck.

## VERTICAL PART
**Origin**
Anterior surface of upper three thoracic and lower three cervical vertebral bodies.

**Insertion**
Transverse processes of fifth and sixth cervical vertebrae (C5, 6).

**Action**
Flexes cervical vertebrae.

**Nerve**
Ventral rami of cervical nerves C2–7.

**Basic functional movement**
Gives smoothness and stability to flexion at the neck.

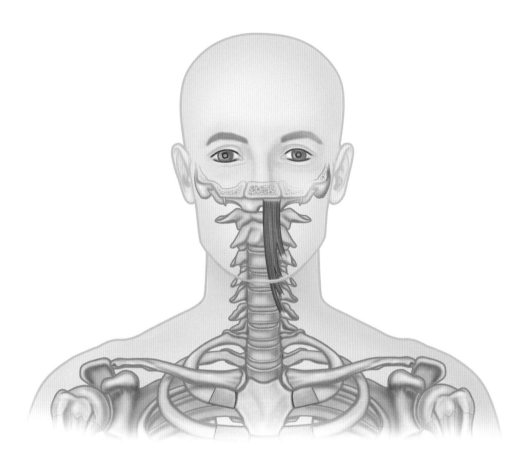

**Latin**, *longus*, long; *capitis*, of the head.

Longus capitis lies anterior to the superior oblique fibers of longus colli.

**Origin**
Transverse processes of third to sixth cervical vertebrae (C3–6).

**Insertion**
Occipital bone anterior to foramen magnum.

**Action**
Flexes head and upper part of cervical spine.

**Nerve**
Ventral rami of cervical nerves C1–3, (C4).

**Basic functional movement**
Gives smoothness and stability to flexion at the head (nodding).

# RECTUS CAPITIS ANTERIOR

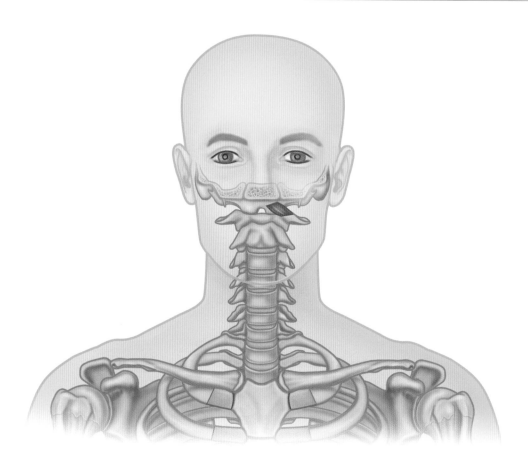

**Latin**, *rectus*, straight; *capitis*, of the head; *anterior*, at the front.

**Origin**
Anterior surface of lateral mass of atlas.

**Insertion**
Basilar part of occipital bone, anterior to occipital condyle (i.e. between occipital condyle and longus capitis).

**Action**
Flexes head upon neck. Holds articular surfaces of atlanto-occipital joint in close apposition during movements.

**Nerve**
Loop between ventral rami of cervical nerves C1, 2.

**Basic functional movement**
Gives smoothness and stability to flexion at the head (nodding).

# RECTUS CAPITIS LATERALIS

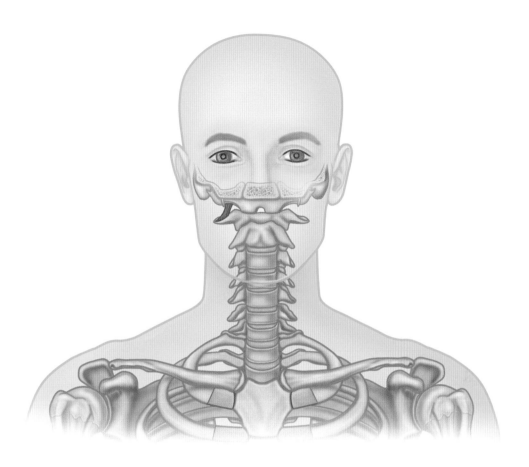

**Latin**, *rectus*, straight; *capitis*, of the head; *lateralis*, relating to the side.

**Origin**
Transverse process of atlas.

**Insertion**
Jugular process of occipital bone.

**Action**
Tilts head laterally to same side. Stabilizes atlanto-occipital joint.

**Nerve**
Loop between ventral rami of cervical nerves C1, 2.

# Lateral Vertebral Muscles

The lateral vertebral muscles of the neck comprise the scalene group (running from the transverse processes of the cervical vertebrae downward to the ribs) and the sternocleidomastoideus.

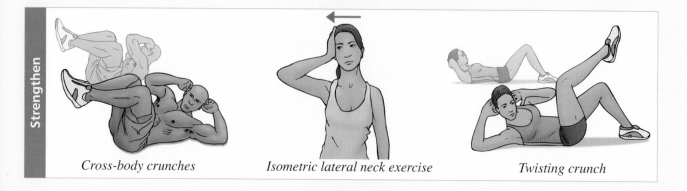

*Cross-body crunches*          *Isometric lateral neck exercise*          *Twisting crunch*

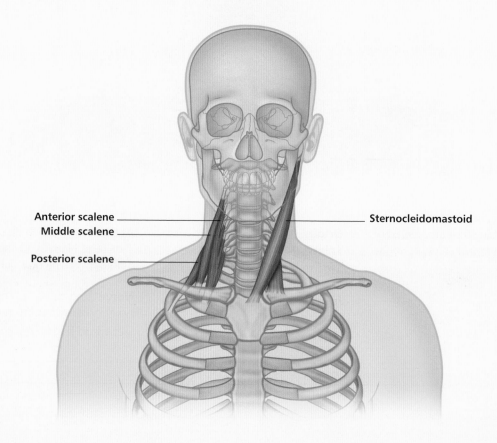

Anterior scalene _____          _____ Sternocleidomastoid
Middle scalene _____
Posterior scalene _____

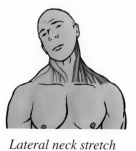 

*Lateral neck stretch*          *Rotating neck stretch*

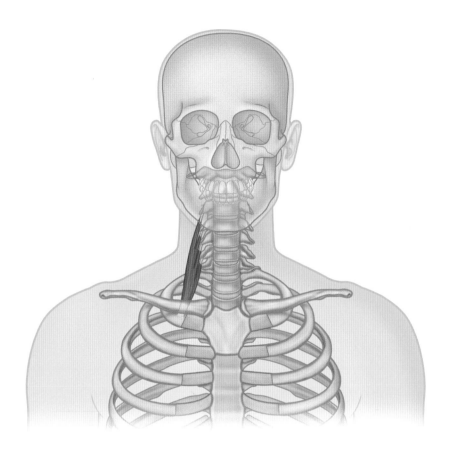

**Greek**, *skalenos*, uneven. **Latin**, *anterior*, at the front.

### Origin
Transverse processes of third to sixth cervical vertebrae (C3–6).

### Insertion
Scalene tubercle on inner border of first rib.

### Action
Acting on both sides: flex neck; raise first rib during active respiratory inhalation.
Acting on one side: laterally flexes and rotates neck.

### Nerve
Ventral rami of cervical nerves C5–7.

### Basic functional movement
Primarily a muscle of inspiration.

### Sports that heavily utilize this muscle
All active sports that require strong respiration (e.g. high-pace running).

### Common problems when muscle is chronically tight/ shortened
Painful conditions of the neck, shoulder, and arm, because hypertonic muscle puts pressure on a bundle of nerves called the brachial plexus as well as on the subclavian artery.

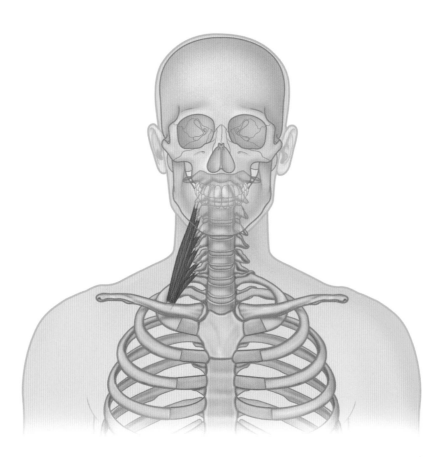

**Greek**, *skalenos*, uneven. **Latin**, *medius*, middle.

### Origin
Posterior tubercles of transverse processes of lower six cervical vertebrae (C2–7).

### Insertion
Upper surface of first rib, behind groove for subclavian artery.

### Action
Acting on both sides: flex neck; raise first rib during active respiratory inhalation.
Acting on one side: laterally flexes and rotates neck.

### Nerve
Ventral rami of cervical nerves C3–8.

### Basic functional movement
Primarily a muscle of inspiration.

### Sports that heavily utilize this muscle
All active sports that require strong respiration (e.g. high-pace running).

### Common problems when muscle is chronically tight/ shortened
Painful conditions of the neck, shoulder, and arm, because hypertonic muscle puts pressure on a bundle of nerves called the brachial plexus as well as on the subclavian artery.

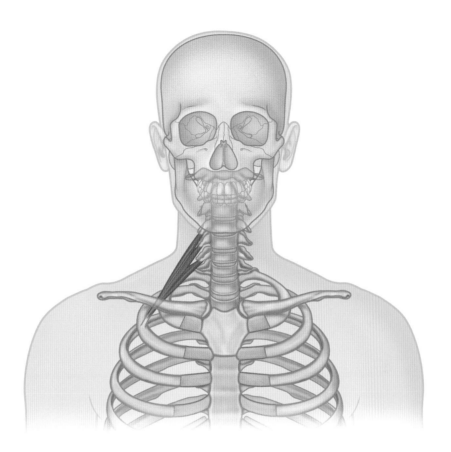

**Greek**, *skalenos*, uneven. **Latin**, *posterior*, at the back.

### Origin
Posterior tubercles of transverse processes of lower two or three cervical vertebrae (C5–7).

### Insertion
Outer surface of second rib.

### Action
Acting on both sides: flex neck; raise second rib during active respiratory inhalation.
Acting on one side: laterally flexes and rotates neck.

### Nerve
Ventral rami of lower cervical nerves C7, 8.

### Basic functional movement
Primarily a muscle of inspiration.

### Sports that heavily utilize this muscle
All active sports that require strong respiration (e.g. high-pace running).

### Common problems when muscle is chronically tight/ shortened
Painful conditions of the neck, shoulder, and arm, because hypertonic muscle puts pressure on a bundle of nerves called the brachial plexus as well as on the subclavian artery.

# STERNOCLEIDOMASTOIDEUS

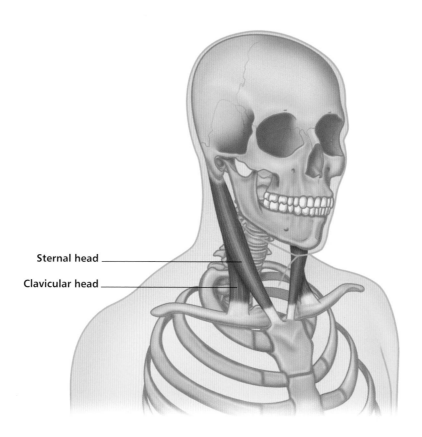

Sternal head

Clavicular head

**Greek**, *sternon*, chest; *kleis*, key; *mastoeides*, breast shaped.

This muscle is a long strap muscle with two heads. It is sometimes injured at birth, and may be partly replaced by fibrous tissue that contracts to produce a torticollis (wry neck).

## Origin
Sternal head: anterior surface of manubrium of sternum.
Clavicular head: upper surface of medial third of clavicle.

## Insertion
Outer surface of mastoid process of temporal bone. Lateral third of superior nuchal line of occipital bone.

## Action
Bilateral contraction: flexes neck and draws head forward, as in raising head from pillow; raises sternum, and consequently ribs, superiorly during deep inhalation. Contraction of one side: tilts head toward same side; rotates head to face opposite side (and also upward as it does so).

## Nerve
Accessory **XI** nerve, with sensory supply for proprioception from cervical nerves C2 and C3.

## Basic functional movement
Examples: turning the head to look over the shoulder, raising the head from a pillow.

## Sports that heavily utilize this muscle
Examples: swimming, rugby scrummage, American football.

## Movements or injuries that may damage this muscle
Extreme whiplash movements.

## Common problems when muscle is chronically tight/ shortened
Headache and neck pain.

# 5 Muscles of the Trunk

The muscles around the spine, and the broader area of the back, are primarily responsible for stabilizing the spinal column and keeping the back in an upright position. The muscles of the back and sides allow the upper body and spine to move in flexion, lateral flexion, extension, hyperextension, and rotation.

**Erector spinae**, also called *sacrospinalis*, comprises three sets of muscles organized in parallel: from lateral to medial these are iliocostalis, longissimus, and spinalis. **Iliocostalis** is the most lateral part of erector spinae and may be subdivided into lumborum, thoracis, and cervicis portions. **Longissimus** is the intermediate part of erector spinae and may be subdivided into thoracis, cervicis, and capitis portions. The **spinalis** is the most medial part of erector spinae and may also be subdivided into thoracis, cervicis, and capitis portions.

The **transversospinalis** muscles are a composite of three small muscle groups situated deep to erector spinae; however, unlike erector spinae, each group lies successively deeper from the surface rather than side by side. From more superficial to deep, the muscle groups are semispinalis, multifidus, and rotatores. Their fibers generally extend upward and medially from transverse processes to higher spinous processes. **Semispinalis** may be subdivided into thoracis, cervicis, and capitis portions. **Multifidus** is the part of the transversospinalis group that lies in the furrow between the spines of the vertebrae and their transverse processes; the muscle lies deep to semispinalis and erector spinae. **Rotatores** are the deepest layer of the transversospinalis group.

**Interspinales** are short and insignificant muscles positioned either side of interspinous ligaments. Like the interspinales, the **intertransversarii** are also short and insignificant muscles; the cervical and thoracic regions encompass intertransversarii anteriores and intertransversarii posteriores, and the lumbar region encompasses intertransversarii laterales and intertransversarii mediales.

The lower **external intercostal** muscles may blend with the fibers of the external oblique, which overlap them, thus effectively forming one continuous sheet of muscle, with the external intercostal fibers seemingly stranded between the ribs. **Internal intercostal** fibers lie deep to, and run obliquely across, the external intercostals. There are eleven external and eleven internal intercostals on each side of the ribcage.

It is interesting to note that the fibrous pericardium is continuous with the central tendon of the **diaphragm** and, in turn, is in direct continuity with psoas major, a structure that has attachments to the vertebral column and the lower limb (lesser trochanter). The diaphragm has both a costal attachment shared with the abdominal obliques and a sternal, or xiphoid, attachment, revealing a fascial continuity with the rectus abdominis. While most textbooks concentrate on describing the psoas as a hip flexor, it is worthwhile giving consideration to its role in breathing and spinal stability. The abdominal obliques should also be regarded as having a direct influence on respiratory function. While muscles may have anatomical individuality, they have little functional individuality. The diaphragm is an essential structure in functional breathing, but compartmentalizing muscle function can mean that we miss synergistic relationships which, if addressed, could improve movement, breathing, and global kinematics.

The muscles of the anterior abdominal wall lie between the ribs and the pelvis, encircling the internal organs, and act to support the trunk, permit movement (primarily flex and rotate the lumbar spine), and support the low back. There are three layers of muscle, with fibers running in the same direction as the corresponding three layers of muscle in the thoracic wall. The deepest layer consists of **transversus abdominis**, whose fibers run approximately horizontally. Transversus abdominis extends around the trunk, to attach into the *thoracolumbar fascia*, a thick connective tissue sheath that helps to stabilize the trunk and pelvis when muscles connecting into it are under tension. The middle layer comprises **obliquus internus abdominis**, whose fibers are crossed by an outermost layer, namely **obliquus externus abdominis**, forming a pattern of fibers resembling a St. Andrew's cross. Overlying these three layers is **rectus abdominis**, which runs vertically, either side of the midline of the abdomen, and is associated with the "six-pack" seen in conditioned athletes. Rectus abdominis is active in trunk flexion, bringing the ribcage closer to the pubic bone and should be trained in a functional manner. Like the other abdominals, it acts as a stabilizing muscle, and also acts as a restraint to hyperextension in the lumbar vertebrae.

**Quadratus lumborum** has fibers running in a crisscross fashion from the twelfth rib above and the first four lumbar vertebrae, to the ilium and iliolumbar ligament below. Its action, while still to be determined by research, is to side bend the trunk, and also to resist the trunk being pulled sideways in the opposite direction.

**Psoas major** runs downward, to be joined by **iliacus**—collectively they are called *iliopsoas*. Together, these muscles act as padding for various abdominal viscera, and leave the abdomen to become the main flexor of the hip joint and stabilizers of the low back. The psoas is to the diaphragm what the gluteus maximus is to the latissimus dorsi. Note that some upper fibers of psoas major may insert by a long tendon into the iliopubic eminence to form **psoas minor**, which has little function and is absent in about forty percent of people. Bilateral contraction of psoas major will increase lumbar lordosis.

# Postvertebral Muscles

The postvertebral muscles are the deepest muscles of the back and run longitudinally on the vertebral column. They are vital in maintaining posture and facilitating movements of the vertebral column. The fibers of the more superficial muscles of this group travel a considerable distance between their origins and insertions, whereas the fibers of the deepest muscles stretch only between one vertebra and the next. Erector spinae, also called *sacrospinalis*, comprises three sets of muscles organized in parallel columns: iliocostalis, longissimus, and spinalis (from lateral to medial).

**Strengthen**

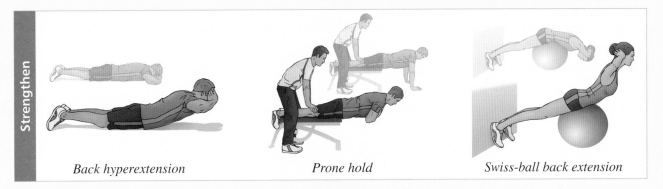

*Back hyperextension*     *Prone hold*     *Swiss-ball back extension*

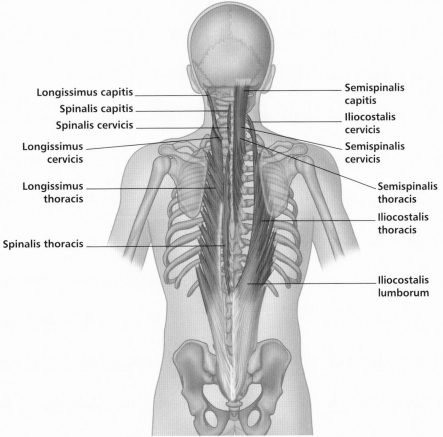

Longissimus capitis

Spinalis capitis

Spinalis cervicis

Longissimus cervicis

Longissimus thoracis

Spinalis thoracis

Semispinalis capitis

Iliocostalis cervicis

Semispinalis cervicis

Semispinalis thoracis

Iliocostalis thoracis

Iliocostalis lumborum

**Self-stretch**

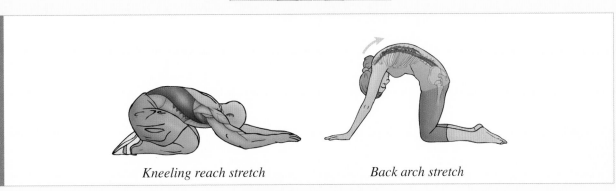

*Kneeling reach stretch*     *Back arch stretch*

# ILIOCOSTALIS LUMBORUM

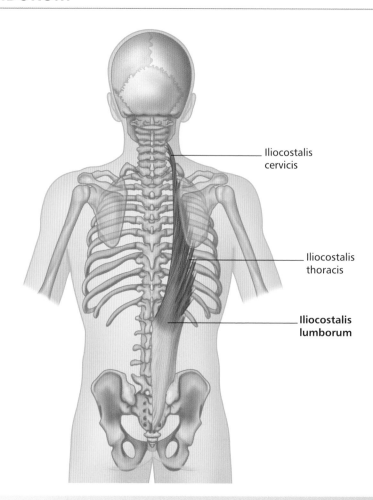

Iliocostalis cervicis

Iliocostalis thoracis

Iliocostalis lumborum

**Latin**, *iliocostalis*, from ilium to rib; *lumborum*, of the loins.

Iliocostalis is the most lateral part of erector spinae and may be subdivided into lumborum, thoracis, and cervicis portions. As a whole, iliocostalis is innervated via the dorsal rami of spinal nerves C4–S5.

### Origin
Lateral and medial sacral crests. Medial part of iliac crests.

### Insertion
Angles of lower six ribs.

### Action
Extends and laterally flexes vertebral column. Helps maintain correct curvature of spine in standing and sitting positions. Steadies vertebral column on pelvis during walking.

### Nerve
Dorsal rami of lumbar nerves.

### Basic functional movement
Keeps the back straight (with correct curvatures).

### Sports that heavily utilize this muscle
Examples: all sports, especially swimming, gymnastics, and wrestling.

### Movements or injuries that may damage this muscle
Lifting without bending the knees or keeping the back erect. Holding an object too far in front of the body when lifting.

# ILIOCOSTALIS THORACIS

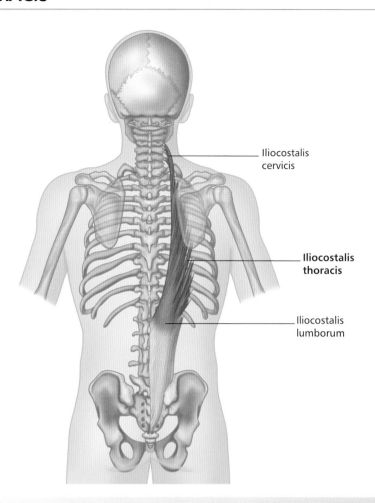

Iliocostalis
cervicis

Iliocostalis
thoracis

Iliocostalis
lumborum

**Latin**, *iliocostalis*, from ilium to rib; *thoracis*, of the chest.

**Origin**
Angles of lower six ribs, medial to iliocostalis lumborum.

**Insertion**
Angles of upper six ribs and transverse process of seventh cervical vertebra (C7).

**Action**
Extends and laterally flexes vertebral column. Helps maintain correct curvature of spine in standing and sitting positions. Rotates ribs for forceful inhalation.

**Nerve**
Dorsal rami of thoracic (intercostal) nerves.

**Basic functional movement**
Keeps the back straight (with correct curvatures).

**Sports that heavily utilize this muscle**
Examples: all sports, especially swimming, gymnastics, and wrestling.

**Movements or injuries that may damage this muscle**
Lifting without bending the knees or keeping the back erect. Holding an object too far in front of the body when lifting.

# ILIOCOSTALIS CERVICIS

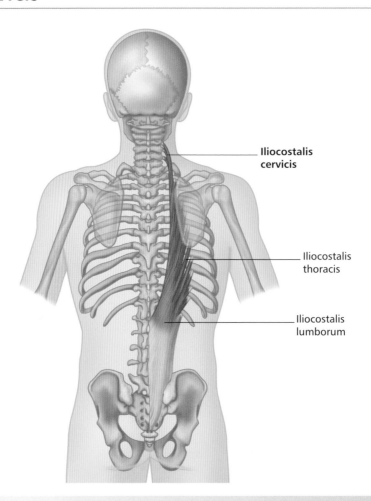

Iliocostalis
cervicis

Iliocostalis
thoracis

Iliocostalis
lumborum

**Latin**, *iliocostalis*, from ilium to rib; *cervicis*, of the neck.

**Origin**
Angles of third to sixth ribs.

**Insertion**
Transverse processes of fourth, fifth, and sixth cervical vertebrae (C4–6).

**Action**
Extends and laterally flexes vertebral column. Helps maintain correct curvature of spine in standing and sitting positions.

**Nerve**
Dorsal rami of cervical nerves.

**Basic functional movement**
Keeps the back straight (with correct curvatures).

**Sports that heavily utilize this muscle**
Examples: all sports, especially swimming, gymnastics, and wrestling.

**Movements or injuries that may damage this muscle**
Lifting without bending the knees or keeping the back erect. Holding an object too far in front of the body when lifting.

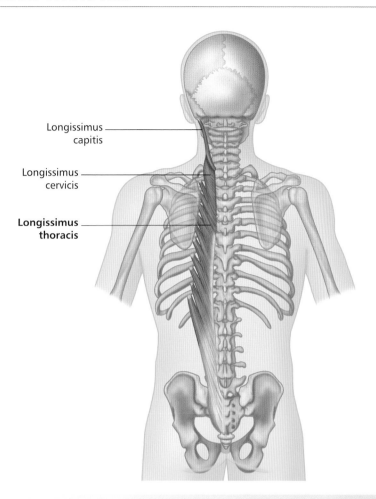

Longissimus capitis

Longissimus cervicis

**Longissimus thoracis**

**Latin**, *longissimus*, longest; *thoracis*, of the chest.

Longissimus is the intermediate part of erector spinae and may be subdivided into thoracis, cervicis, and capitis portions. As a whole, longissimus is innervated via the dorsal rami of spinal nerves C1–S1.

### Origin
Lateral and medial sacral crests. Spinous processes and supraspinal ligament of all lumbar vertebrae (L1–5) and eleventh and twelfth thoracic vertebrae (T11–12). Medial part of iliac crest.

### Insertion
Transverse processes of all thoracic vertebrae (T1–12). Area between tubercles and angles of lower nine or ten ribs.

### Action
Extends and laterally flexes vertebral column. Helps maintain correct curvature of spine in standing and sitting positions. Rotates ribs for forceful inhalation. Steadies vertebral column on pelvis during walking.

### Nerve
Dorsal rami of lumbar and thoracic nerves.

### Basic functional movement
Keeps the back straight (with correct curvatures).

### Sports that heavily utilize this muscle
Examples: all sports, especially swimming, gymnastics, and wrestling.

### Movements or injuries that may damage this muscle
Lifting without bending the knees or keeping the back erect. Holding an object too far in front of the body when lifting.

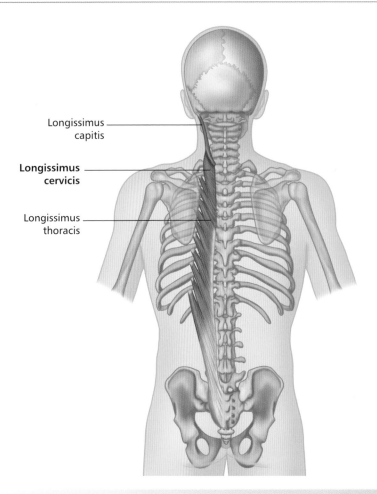

Longissimus
capitis

**Longissimus
cervicis**

Longissimus
thoracis

**Latin**, *longissimus*, longest;
*cervicis*, of the neck.

**Origin**
Transverse processes of upper four
or five thoracic vertebrae (T1–5).

**Insertion**
Transverse processes of second to
sixth cervical vertebrae (C2–6).

**Action**
Extends and laterally flexes upper
vertebral column. Helps maintain
correct curvature of thoracic and
cervical spine in standing and
sitting positions.

**Nerve**
Dorsal rami of lower cervical and
upper thoracic nerves.

**Basic functional movement**
Keeps the upper back and neck
straight (with correct curvatures).

**Sports that heavily utilize this
muscle**
Examples: all sports, especially
swimming, gymnastics, and
wrestling.

**Movements or injuries that
may damage this muscle**
Lifting without bending the knees
or keeping the back erect. Holding
an object too far in front of the
body when lifting.

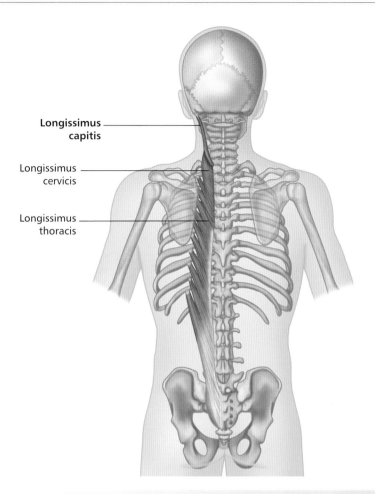

Longissimus capitis

Longissimus cervicis

Longissimus thoracis

**Latin**, *longissimus*, longest; *capitis*, of the head.

### Origin
Transverse processes of upper five thoracic vertebrae (T1–5). Articular processes of lower three cervical vertebrae (C5–7).

### Insertion
Posterior part of mastoid process of temporal bone.

### Action
Extends and rotates head. Helps maintain correct curvature of thoracic and cervical spine in standing and sitting positions.

### Nerve
Dorsal rami of middle and lower cervical nerves.

### Basic functional movement
Keeps the upper back straight (with correct curvatures).

### Sports that heavily utilize this muscle
Examples: all sports, especially swimming, gymnastics, and wrestling.

### Movements or injuries that may damage this muscle
Lifting without bending the knees or keeping the back erect. Holding an object too far in front of the body when lifting.

# SPINALIS THORACIS

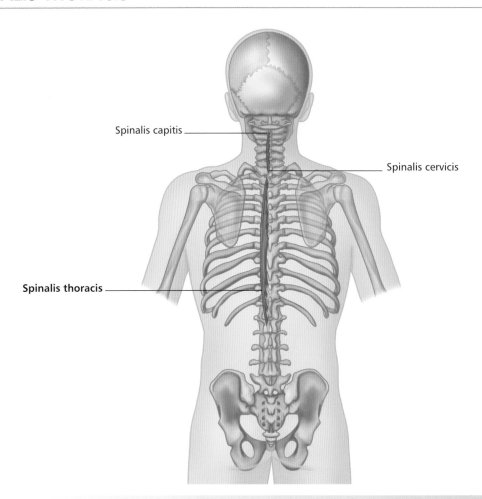

Spinalis capitis

Spinalis cervicis

Spinalis thoracis

**Latin**, *spinalis*, relating to the spine; *thoracis*, of the chest.

The spinalis is the most medial part of erector spinae and may be subdivided into thoracis, cervicis, and capitis portions. As a whole, the spinalis is innervated via the dorsal rami of spinal nerves C2–L3.

### Origin
Spinous processes of lower two thoracic vertebrae (T11–12) and upper two lumbar vertebrae (L1–2).

### Insertion
Spinous processes of upper eight thoracic vertebrae (T1–8).

### Action
Extends vertebral column. Helps maintain correct curvature of spine in standing and sitting positions.

### Nerve
Dorsal rami of spinal nerves.

### Basic functional movement
Keeps the back straight (with correct curvatures).

### Sports that heavily utilize this muscle
Examples: all sports, especially swimming, gymnastics, and wrestling.

### Movements or injuries that may damage this muscle
Lifting without bending the knees or keeping the back erect. Holding an object too far in front of the body when lifting.

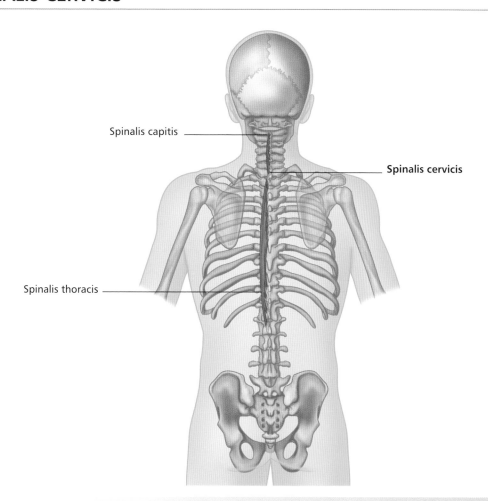

Spinalis capitis

Spinalis cervicis

Spinalis thoracis

**Latin**, *spinalis*, relating to the spine; *cervicis*, of the neck.

### Origin
Ligamentum nuchae. Spinous process of seventh cervical vertebra (C7).

### Insertion
Spinous process of axis.

### Action
Extends vertebral column. Helps maintain correct curvature of cervical spine in standing and sitting positions.

### Nerve
Dorsal rami of cervical nerves.

### Basic functional movement
Keeps the neck straight (with correct curvatures).

### Sports that heavily utilize this muscle
Examples: all sports, especially swimming, gymnastics, and wrestling.

### Movements or injuries that may damage this muscle
Lifting without bending the knees or keeping the back erect. Holding an object too far in front of the body when lifting.

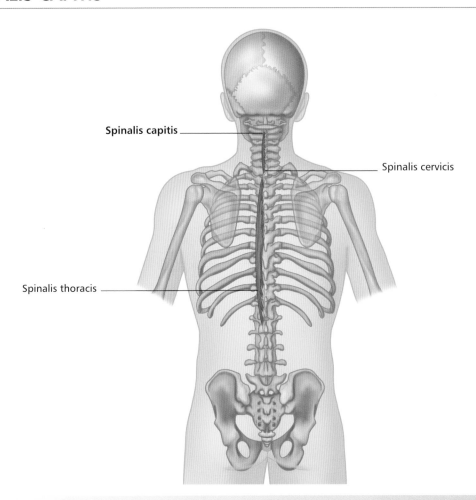

Spinalis capitis

Spinalis cervicis

Spinalis thoracis

**Latin**, *spinalis*, relating to the spine; *capitis*, of the head.

Medial part of semispinalis capitis.

## Origin
Lateral aspect of spinous process of C7.

## Insertion
Medial portion of nuchal line, between superior and inferior lines.

## Action
Extends head and vertebrae.

## Nerve
Dorsal rami of cervical nerves C1–3.

## Sports that heavily utilize this muscle
Examples: rugby scrums, American football, wrestling, swimming.

## Movements or injuries that may damage this muscle
Whiplash injuries.

## Common problems when muscle is chronically tight/shortened
Headache and neck pain.

# SPLENIUS CAPITIS

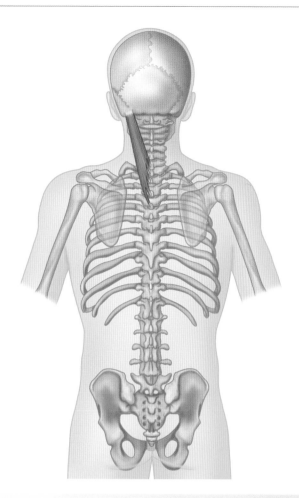

**Greek**, *splenion*, bandage. **Latin**, *capitis*, of the head.

### Origin
Lower part of ligamentum nuchae. Spinous processes of seventh cervical vertebra (C7) and upper three or four thoracic vertebrae (T1–4).

### Insertion
Posterior aspect of mastoid process of temporal bone. Lateral part of superior nuchal line, deep to attachment of sternocleidomastoideus.

### Action
Acting on both sides: extend head and neck.
Acting on one side: laterally flexes neck; rotates head to same side as contracting muscle.

### Nerve
Dorsal rami of middle and lower cervical nerves.

### Basic functional movement
Example: looking up, or turning the head to look behind.

### Sports that heavily utilize this muscle
Examples: rugby scrums, American football, wrestling, swimming.

### Movements or injuries that may damage this muscle
Whiplash injuries.

### Common problems when muscle is chronically tight/shortened
Headache and neck pain.

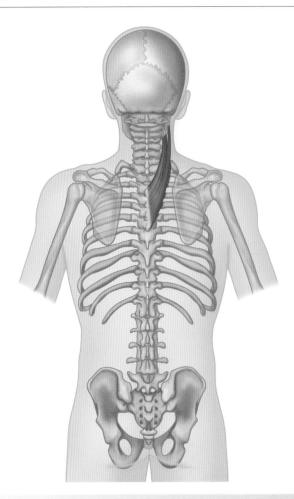

**Greek**, *splenion*, bandage. **Latin**, *cervicis*, of the neck.

**Origin**
Spinous processes of third to sixth thoracic vertebrae (T3–6).

**Insertion**
Posterior tubercles of transverse processes of upper two or three cervical vertebrae (C1–3).

**Action**
Acting on both sides: extend head and neck.
Acting on one side: laterally flexes neck; rotates head to same side as contracting muscle.

**Nerve**
Dorsal rami of middle and lower cervical nerves.

**Basic functional movement**
Example: looking up, or turning the head to look behind.

**Sports that heavily utilize this muscle**
Examples: rugby scrums, American football, wrestling, swimming.

**Movements or injuries that may damage this muscle**
Whiplash injuries.

**Common problems when muscle is chronically tight/shortened**
Headache and neck pain.

# Transversospinalis Muscles

The transversospinalis muscles are a composite of three small muscle groups situated deep to erector spinae; however, unlike erector spinae, each group lies successively deeper from the surface, rather than side by side. From more superficial to deep, the muscle groups are semispinalis, multifidus, and rotatores. Their fibers generally extend upward and medially from transverse processes to higher spinous processes.

**Strengthen**

*Back hyperextension*          *Prone hold*          *Swiss-ball back extension*

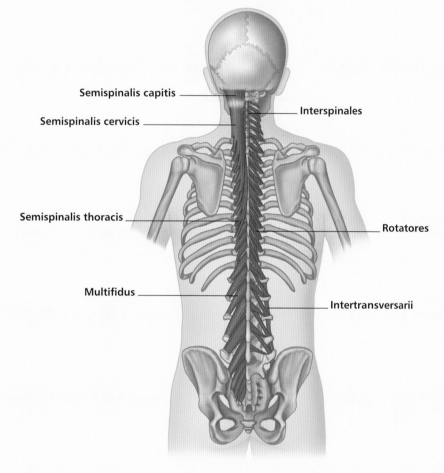

Semispinalis capitis — 
Semispinalis cervicis — 
— Interspinales
Semispinalis thoracis — 
— Rotatores
Multifidus — 
— Intertransversarii

**Self-stretch**

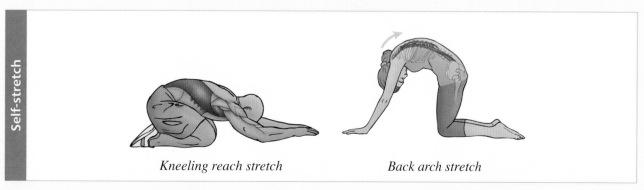

*Kneeling reach stretch*          *Back arch stretch*

123

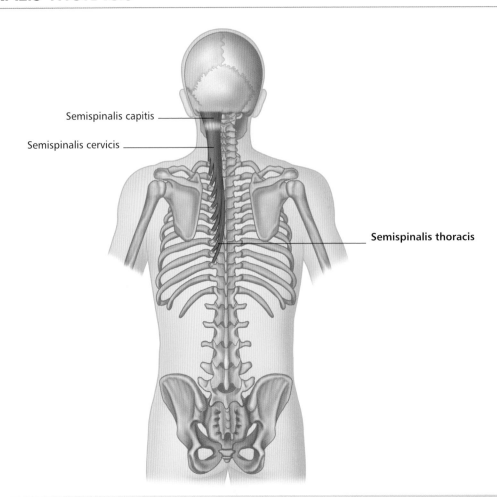

Semispinalis capitis

Semispinalis cervicis

Semispinalis thoracis

**Latin**, *semispinalis*, half-spinal; *thoracis*, of the chest.

## Origin
Transverse processes of sixth to tenth thoracic vertebrae (T6–10).

## Insertion
Spinous processes of lower two cervical and upper four thoracic vertebrae (C6–T4).

### Action
Extends thoracic and cervical parts of vertebral column. Assists in rotation of thoracic and cervical vertebrae.

### Nerve
Dorsal rami of thoracic and cervical spinal nerves.

### Basic functional movement
Example: looking up, or turning the head to look behind.

### Sports that heavily utilize this muscle
Examples: rugby scrums, American football, wrestling, swimming.

### Movements or injuries that may damage this muscle
Whiplash injuries.

# SEMISPINALIS CERVICIS

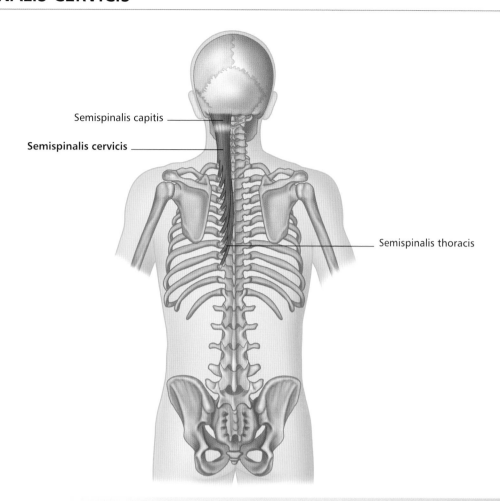

Semispinalis capitis

**Semispinalis cervicis**

Semispinalis thoracis

**Latin**, *semispinalis*, half-spinal; *cervicis*, of the neck.

**Origin**
Transverse processes of upper five or six thoracic vertebrae (T1–6).

**Insertion**
Spinous processes of second to fifth cervical vertebrae (C2–5).

**Action**
Extends thoracic and cervical parts of vertebral column. Assists in rotation of thoracic and cervical vertebrae.

**Nerve**
Dorsal rami of thoracic and cervical spinal nerves.

**Basic functional movement**
Example: looking up, or turning the head to look behind.

**Sports that heavily utilize this muscle**
Examples: rugby scrums, American football, wrestling, swimming.

**Movements or injuries that may damage this muscle**
Whiplash injuries.

# SEMISPINALIS CAPITIS

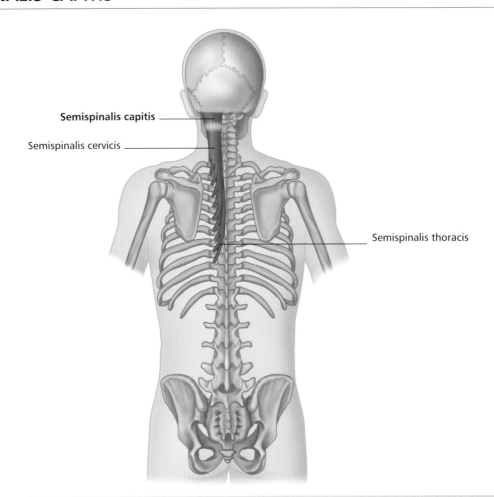

Semispinalis capitis

Semispinalis cervicis

Semispinalis thoracis

**Latin**, *semispinalis*, half-spinal; *capitis*, of the head.

Medial part is spinalis capitis.

### Origin
Transverse processes of lower four cervical and upper six or seven thoracic vertebrae (C4–T7).

### Insertion
Between superior and inferior nuchal lines of occipital bone.

### Action
Most powerful extensor of head. Assists in rotation of head.

### Nerve
Dorsal rami of cervical nerves.

### Basic functional movement
Example: looking up, or turning the head to look behind.

### Sports that heavily utilize this muscle
Examples: rugby scrums, American football, wrestling, swimming.

### Movements or injuries that may damage this muscle
Whiplash injuries.

# MULTIFIDUS

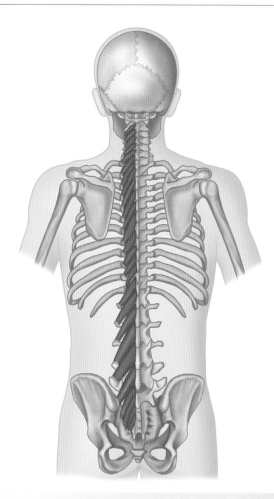

**Latin**, *multi*, many; *findere*, to split.

This muscle is the part of the transversospinalis group that lies in the furrow between the spines of the vertebrae and their transverse processes. Multifidus lies deep to semispinalis and erector spinae.

### Origin
Posterior surface of sacrum, between sacral foramina and posterior superior iliac spine. Mammillary processes (posterior borders of superior articular processes) of all lumbar vertebrae. Transverse processes of all thoracic vertebrae. Articular processes of lower four cervical vertebrae.

### Insertion
Parts insert into the spinous processes of the two to four vertebrae superior to the origin; overall this includes the spinous processes of all the vertebrae from fifth lumbar up to axis (L5–C2).

### Action
Protects vertebral joints from movements produced by the more powerful superficial prime movers. Extension, lateral flexion, and rotation of vertebral column.

### Nerve
Dorsal rami of spinal nerves.

### Basic functional movement
Helps maintain good posture and spinal stability during standing, sitting, and all movements.

### Movements or injuries that may damage this muscle
Lifting without bending the knees or keeping the back erect. Holding an object too far in front of the body when lifting.

# ROTATORES

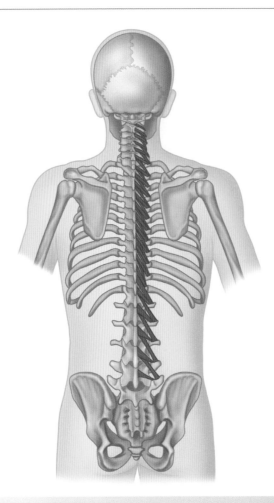

**Latin**, *rota*, wheel.

These small muscles are the deepest layer of the transversospinalis group.

**Origin**
Transverse process of each vertebra.

**Insertion**
Base of spinous process of adjoining vertebra above.

**Action**
Rotate and assist in extension of vertebral column.

**Nerve**
Dorsal rami of spinal nerves.

**Basic functional movement**
Help maintain good posture and spinal stability during standing, sitting, and all movements.

**Movements or injuries that may damage these muscles**
Lifting without bending the knees or keeping the back erect. Holding an object too far in front of the body when lifting.

# INTERSPINALES

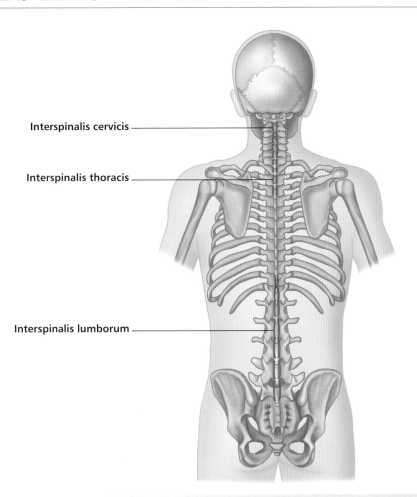

Interspinalis cervicis

Interspinalis thoracis

Interspinalis lumborum

**Latin**, *inter*, between; *spinalis*, relating to the spine.

Short and insignificant muscles positioned either side of interspinous ligament.

### Origin/Insertion
Extend from one spinous process (origin) to the next one above (insertion), throughout the vertebral column. These muscles are most developed in cervical and lumbar regions, and may be absent in thoracic region.

### Action
Act as extensile ligaments. Weakly extend vertebral column.

### Nerve
Dorsal rami of spinal nerves.

# INTERTRANSVERSARII ANTERIORES

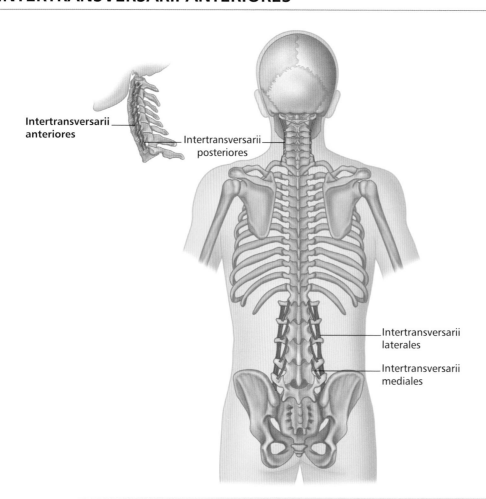

Intertransversarii anteriores

Intertransversarii posteriores

Intertransversarii laterales

Intertransversarii mediales

Like the interspinales, the intertransversarii are also short and insignificant muscles. The cervical and thoracic regions encompass the anterior and posterior intertransversarii, and the lumbar region encompasses the lateral and medial intertransversarii.

**Latin**, *inter*, between; *transversus*, across, crosswise; *anterior*, at the front.

**Origin**
Anterior tubercle of transverse processes of vertebrae from first thoracic to axis (T1–C2).

**Insertion**
Anterior tubercle of adjacent vertebra above.

**Action**
Slightly assist lateral flexion of cervical vertebrae. Act as extensile ligaments.

**Nerve**
Ventral rami of spinal nerves.

# INTERTRANSVERSARII POSTERIORES

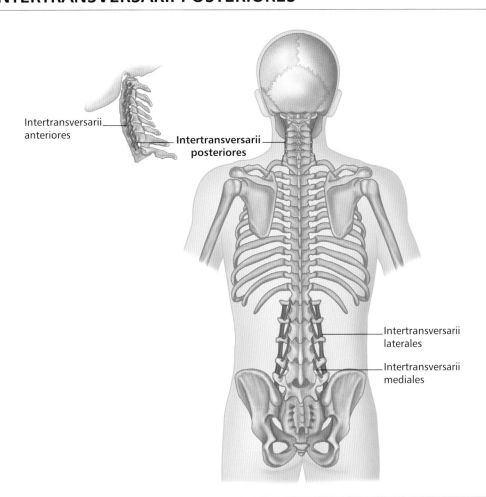

Intertransversarii anteriores

**Intertransversarii posteriores**

Intertransversarii laterales

Intertransversarii mediales

**Latin**, *inter*, between; *transversus*, across, crosswise; *posterior*, at the back.

## Origin
Posterior tubercle of transverse processes of vertebrae from first thoracic to axis (T1–C2). Transverse processes of first lumbar to eleventh thoracic vertebrae (L1–T11).

## Insertion
Transverse process of adjacent vertebra above (posterior tubercles in cervical region).

## Action
In cervical region, slightly assist lateral flexion of cervical vertebrae. Act as extensile ligaments.

## Nerve
Ventral rami of spinal nerves.

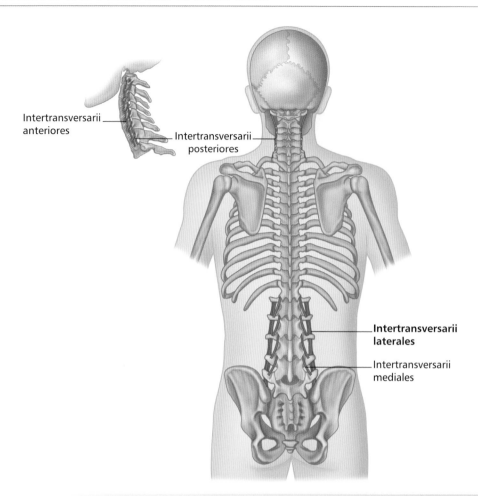

Intertransversarii
anteriores

Intertransversarii
posteriores

**Intertransversarii
laterales**

Intertransversarii
mediales

**Latin**, *inter*, between; *transversus*, across, crosswise; *lateralis*, relating to the side.

**Origin**
Transverse processes of lumbar vertebrae.

**Insertion**
Transverse process of adjacent vertebra above.

**Action**
Slightly assist lateral flexion of lumbar vertebrae. Act as extensile ligaments.

**Nerve**
Ventral rami of spinal nerves.

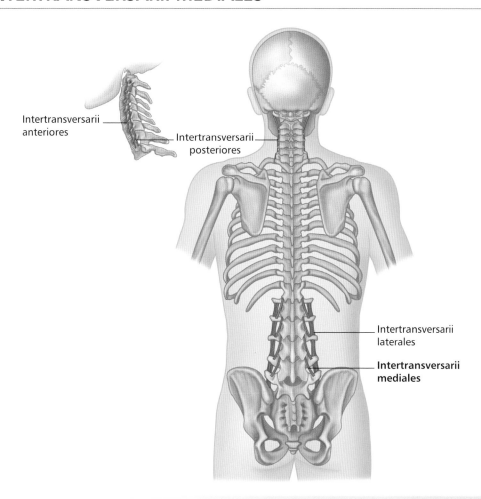

Intertransversarii
anteriores

Intertransversarii
posteriores

Intertransversarii
laterales

**Intertransversarii
mediales**

**Latin**, *inter*, between; *transversus*, across, crosswise; *medialis*, relating to the middle.

**Origin**
Mammillary process (posterior border of superior articular process) of each lumbar vertebra.

**Insertion**
Accessory process of adjacent lumbar vertebra above.

**Action**
Slightly assist lateral flexion of lumbar vertebrae. Act as extensile ligaments.

**Nerve**
Dorsal rami of spinal nerves.

# Postvertebral Muscles—Suboccipital Group

The suboccipital group of muscles lies deep in the neck, anterior to semispinalis capitis, longissimus capitis, and splenius capitis. The muscle group encloses a triangular space known as the *suboccipital triangle*.

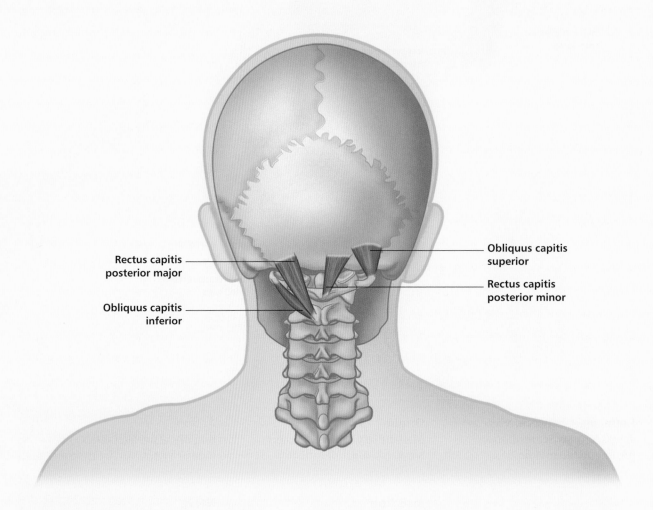

Rectus capitis posterior major

Obliquus capitis inferior

Obliquus capitis superior

Rectus capitis posterior minor

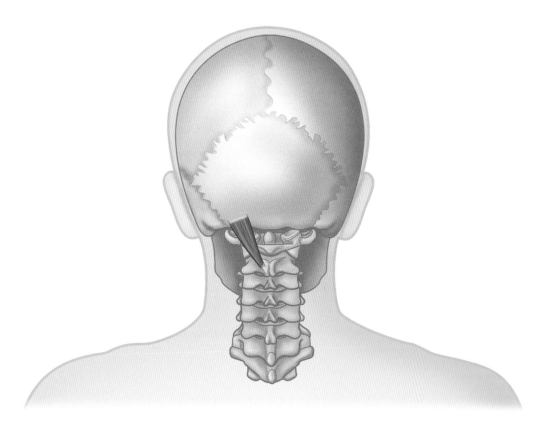

**Latin**, *rectus*, straight; *capitis*, of the head; *posterior*, at the back; *major*, larger.

**Origin**
Spinous process of axis.

**Insertion**
Below lateral portion of inferior nuchal line of occipital bone.

**Action**
Extends head. Rotates head to same side.

**Nerve**
Suboccipital nerve (dorsal ramus of first cervical nerve C1).

**Basic functional movement**
Helps smooth and stabilize the act of looking upward and over the shoulder.

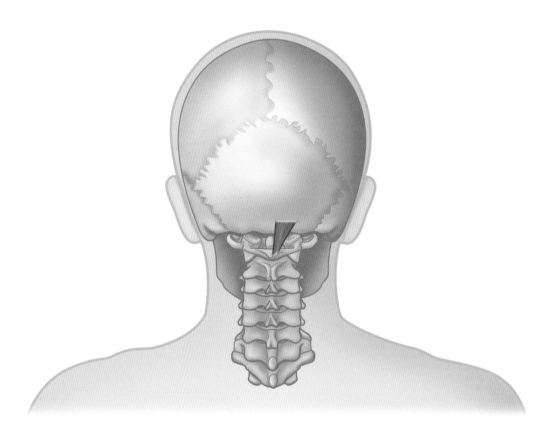

**Latin**, *rectus*, straight; *capitis*, of the head; *posterior*, at the back; *minor*, smaller.

**Origin**
Posterior tubercle of atlas.

**Insertion**
Medial portion of inferior nuchal line of occipital bone.

**Action**
Extends head.

**Nerve**
Suboccipital nerve (dorsal ramus of first cervical nerve C1).

**Basic functional movement**
Helps smooth and stabilize the act of looking upward.

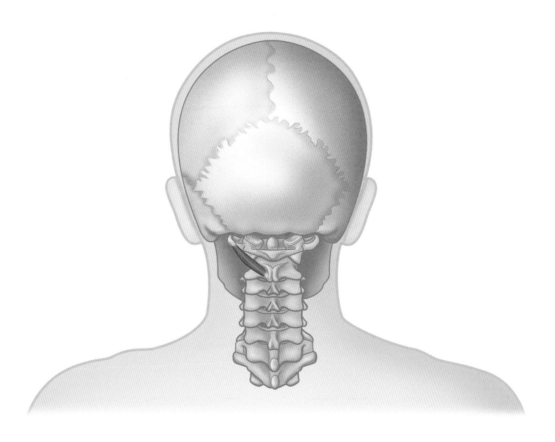

**Latin**, *obliquus*, diagonal, slanted; *capitis*, of the head; *inferior*, lower.

**Origin**
Spinous process of axis.

**Insertion**
Transverse process of atlas.

**Action**
Rotates atlas upon axis, thereby rotating head to same side.

**Nerve**
Suboccipital nerve (dorsal ramus of first cervical nerve C1).

**Basic functional movement**
Gives stability to the head when it is turned.

# OBLIQUUS CAPITIS SUPERIOR

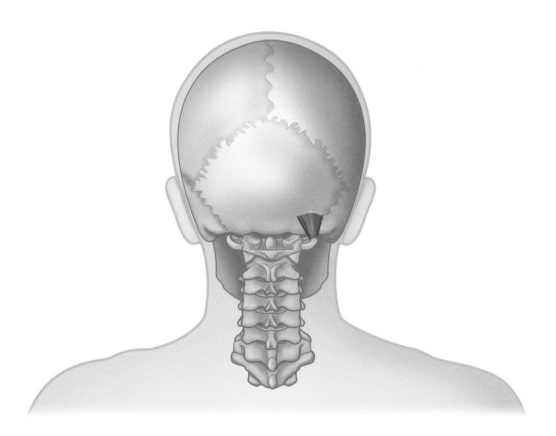

**Latin**, *obliquus*, diagonal, slanted; *capitis*, of the head; *superior*, upper.

**Origin**
Transverse process of atlas.

**Insertion**
Area between inferior and superior nuchal lines on occipital bone.

**Action**
Extends head.

**Nerve**
Suboccipital nerve (dorsal ramus of first cervical nerve C1).

**Basic functional movement**
Helps smooth and stabilize the act of looking upward.

# Muscles of the Thorax

The muscles in this section are small muscles, all primarily concerned with movements of the ribs.

*Cross-body crunches*      *Twisting crunch*      *Weighted seated twist*

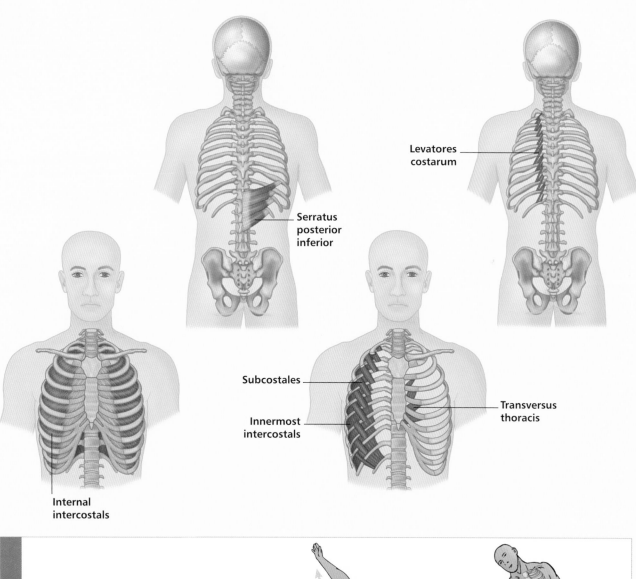

Levatores costarum

Serratus posterior inferior

Subcostales

Innermost intercostals

Transversus thoracis

Internal intercostals

*Back arch stretch*      *Kneeling back rotation stretch*      *Lateral side stretch*

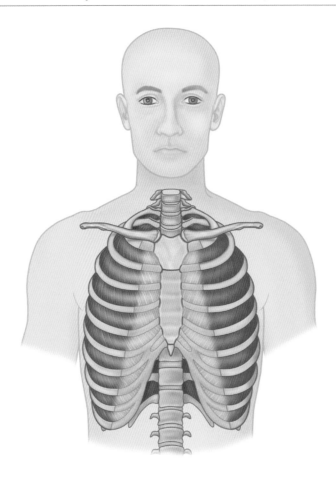

**Latin**, *inter*, between; *costa*, rib; *externi*, external.

The lower external intercostal muscles may blend with the fibers of the external oblique, which overlap them, thus effectively forming one continuous sheet of muscle, with the external intercostal fibers seemingly stranded between the ribs. There are eleven external intercostals on each side of the ribcage.

**Origin**
Lower border of a rib.

**Insertion**
Upper border of rib below (fibers run obliquely forward and downward).

**Action**
Contract to stabilize ribcage during various movements of trunk. May elevate ribs during inspiration, thus increasing volume of thoracic cavity (although this action is disputed). Prevent intercostal space from bulging out or sucking in during respiration.

**Nerve**
The corresponding intercostal nerves.

**Sports that heavily utilize these muscles**
All very active sports.

**Common problems when muscles are chronically tight/ shortened (spastic)**
Kyphosis (rounded back) and depressed chest.

# INTERCOSTALES INTERNI (Internal Intercostals)

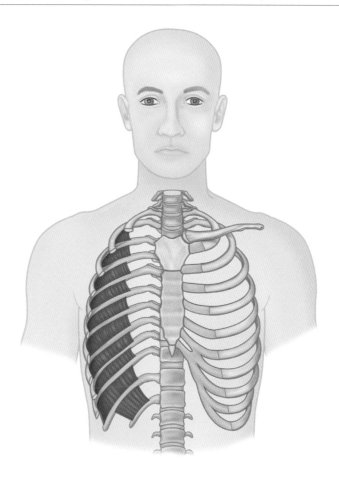

**Latin**, *inter*, between; *costalis*, relating to the ribs; *interni*, internal.

Internal intercostal fibers lie deep to, and run obliquely across, the external intercostals. There are eleven internal intercostals on each side of the ribcage.

**Origin**
Upper border of a rib and costal cartilage.

**Insertion**
Lower border of rib above (fibers run obliquely forward and upward, toward the costal cartilage).

**Action**
Contract to stabilize ribcage during various movements of trunk. May draw adjacent ribs together during forced expiration, thus decreasing volume of thoracic cavity (although this action is disputed). Prevent intercostal space from bulging out or sucking in during respiration.

**Nerve**
The corresponding intercostal nerves.

**Sports that heavily utilize these muscles**
All very active sports.

**Common problems when muscles are chronically tight/ shortened (spastic)**
Kyphosis (rounded back) and depressed chest.

# INTERCOSTALES INTIMI (Innermost Intercostals)

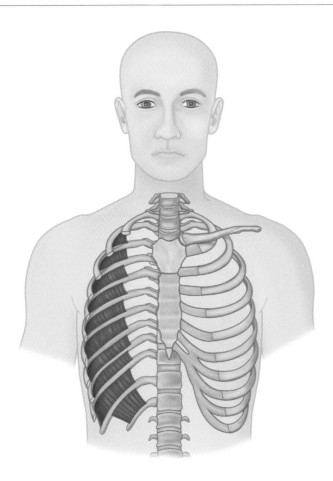

**Latin**, *inter*, between; *costalis*, relating to the ribs; *intimo*, innermost part.

These muscles are variable layers of fibers that run in the same direction as, but deep to, the internal intercostals. They are separated from the internal intercostals by the intercostal nerves and vessels.

**Origin**
Superior border of each rib.

**Insertion**
Inferior border of the preceding rib.

**Action**
While the action of the innermost intercostals is unknown, it is accepted that they act to fix the position of the ribs during respiration.

**Nerve**
Corresponding intercostal nerves.

# SUBCOSTALES

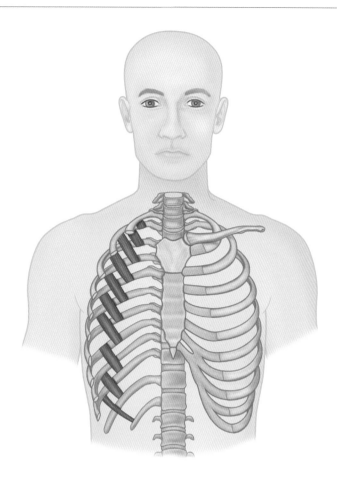

**Latin**, *sub*, under; *costalis*, relating to the ribs.

Positioned deep to the lower internal intercostals, the subcostales fibers run in the same direction as the innermost intercostal muscles and may be continuous with them. Subcostales, transversus thoracis, and the innermost intercostal muscles make up the deepest intercostal muscle layer.

**Origin**
Inner surface of each lower rib near its angle.

**Insertion**
Fibers run obliquely and medially into the inner surface of second or third rib below.

**Action**
Contract to stabilize ribcage during various movements of trunk. May draw adjacent ribs together during forced expiration, thus decreasing volume of thoracic cavity (although this action is disputed).

**Nerve**
The corresponding intercostal nerves.

# TRANSVERSUS THORACIS

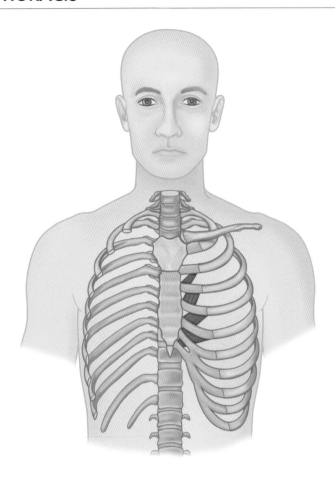

**Latin**, *transversus*, across, crosswise; *thoracis*, of the chest.

Situated deep to the internal intercostals.

**Origin**
Posterior surface of xiphoid process and body of sternum.

**Insertion**
Inner surfaces of costal cartilages of second to sixth ribs.

**Action**
Draws costal cartilages downward, contributing to forceful exhalation.

**Nerve**
The corresponding intercostal nerves.

**Basic functional movement**
Example: blowing out a stubborn flame.

# LEVATORES COSTARUM

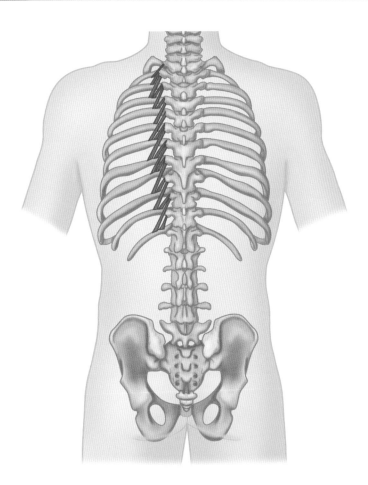

*Posterior view.*

**Latin**, *levare*, to lift; *costarum*, of the ribs.

Small, relatively insignificant muscles.

**Origin**
Transverse processes of seventh cervical to eleventh thoracic vertebrae inclusive (C7–T11).

**Insertion**
Laterally downward to external surface of rib below, between tubercle and angle.

**Action**
Raise the ribs. May very slightly assist lateral flexion and rotation of vertebral column.

**Nerve**
Ventral rami of thoracic spinal nerves.

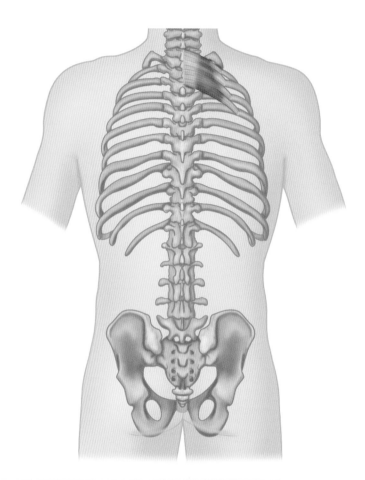

*Posterior view.*

**Latin**, *serratus*, serrated; *posterior*, at the back; *superior*, upper.

Lies deep to the rhomboid muscles.

**Origin**
Lower part of ligamentum nuchae. Spinous processes of seventh cervical vertebra (C7) and upper three or four thoracic vertebrae (T1–4).

**Insertion**
Upper borders of second to fifth ribs, lateral to their angles.

**Action**
Raises upper ribs (probably during forced inhalation).

**Nerve**
Intercostal nerves T2, 3, 4

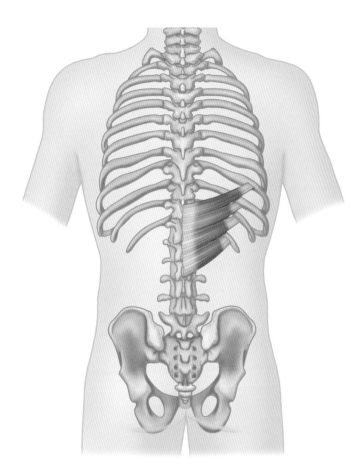

*Posterior view.*

**Latin**, *serratus*, serrated; *posterior*, at the back; *inferior*, lower.

### Origin
Thoracolumbar fascia, at its attachment to spinous processes of lower two thoracic vertebrae (T11–12) and upper two or three lumbar vertebrae (L1–3).

### Insertion
Lower borders of last four ribs.

### Action
May help draw lower ribs downward and backward, resisting the pull of the diaphragm.

### Nerve
Intercostal nerves T9, 10, 11.

# DIAPHRAGM

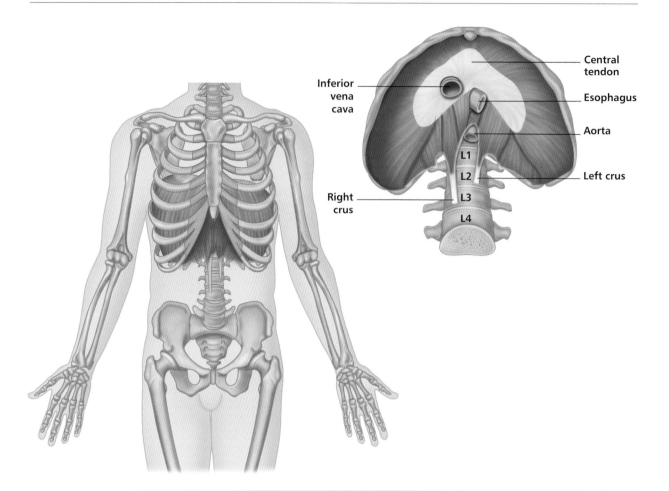

Labels on diagram:
- Inferior vena cava
- Right crus
- Central tendon
- Esophagus
- Aorta
- Left crus
- L1
- L2
- L3
- L4

**Greek**, *dia*, across; *phragma*, partition, wall.

The left and right crus are two tendinous structures that extend below the diaphragm to the vertebral column. Together they act as a tether to aid muscular contraction.

## Origin
Sternal portion: back of xiphoid process.
Costal portion: inner surfaces of lower six ribs and their costal cartilages.
Lumbar portion: upper two or three lumbar vertebrae (L1–3). Medial and lateral lumbocostal arches (also known as the *medial and lateral arcuate ligaments*).

## Insertion
All fibers converge and attach onto a central tendon, i.e. this muscle inserts upon itself.

## Action
Forms floor of thoracic cavity. Pulls central tendon downward during inhalation, thereby increasing volume of thoracic cavity.

## Nerve
Phrenic nerve (ventral rami) C3, **4**, 5.

## Basic functional movement
Produces about sixty percent of breathing capacity.

## Sports that heavily utilize this muscle
All physically demanding sports.

# Muscles of the Anterior Abdominal Wall

The anterior abdominal wall has three layers of muscle, with fibers running in the same direction as the corresponding three layers of muscle in the thoracic wall. The deepest layer consists of transversus abdominis, whose fibers run approximately horizontally. The middle layer comprises obliquus internus abdominis, whose fibers are crossed by an outermost layer, namely obliquus externus abdominis, forming a pattern of fibers resembling a St. Andrew's cross. Overlying these three layers is rectus abdominis, which runs vertically, either side of the midline of the abdomen.

**Strengthen**

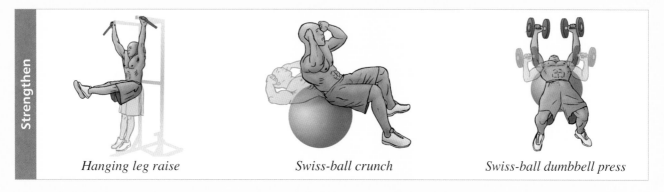

*Hanging leg raise*          *Swiss-ball crunch*          *Swiss-ball dumbbell press*

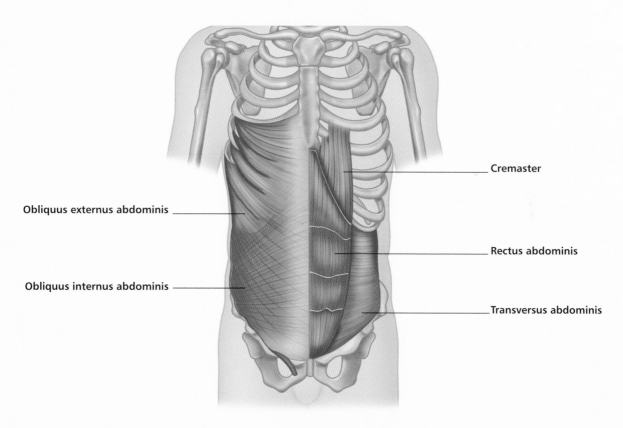

Cremaster

Obliquus externus abdominis

Rectus abdominis

Obliquus internus abdominis

Transversus abdominis

**Self-stretch**

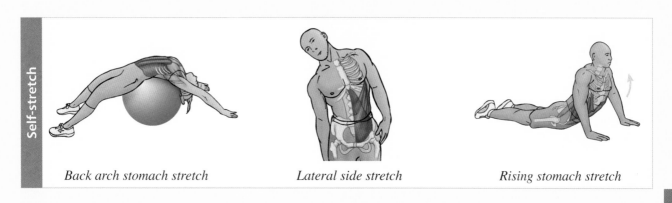

*Back arch stomach stretch*          *Lateral side stretch*          *Rising stomach stretch*

# OBLIQUUS EXTERNUS ABDOMINIS (External Oblique)

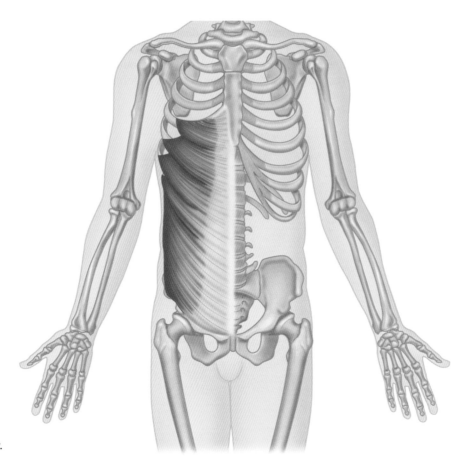

*Anterior view.*

**Latin**, *obliquus*, diagonal, slanted; *externus*, external; *abdominis*, of the belly/stomach.

The posterior fibers of the external oblique are usually overlapped by latissimus dorsi, but in some cases there is a space between the two, known as the *lumbar triangle*, situated just above the iliac crest. The lumbar triangle is a weak point in the abdominal wall.

## Origin
Anterior fibers: outer surfaces of ribs five through eight, interdigitating with serratus anterior.
Lateral fibers: outer surface of ninth rib, interdigitating with serratus anterior; and outer surfaces of tenth, eleventh, and twelfth ribs, interdigitating with latissimus dorsi.

## Insertion
Anterior fibers: into a broad, flat abdominal aponeurosis that terminates in the linea alba, a tendinous raphe extending from xiphoid process.
Lateral fibers: as the inguinal ligament, into anterior superior iliac spine and pubic tubercle, and into lip of anterior one-half of iliac crest.

## Action
Compresses abdomen, helping to support abdominal viscera against pull of gravity. Contraction of one side alone bends trunk laterally to that side and rotates it to the opposite side.

## Nerve
Ventral rami of thoracic nerves T5–12.

## Basic functional movement
Example: digging with a shovel.

## Sports that heavily utilize this muscle
Examples: gymnastics, rowing, rugby.

## Common problems when muscle is weak
Injury to the lumbar spine, because abdominal muscle tone contributes to stability of this region.

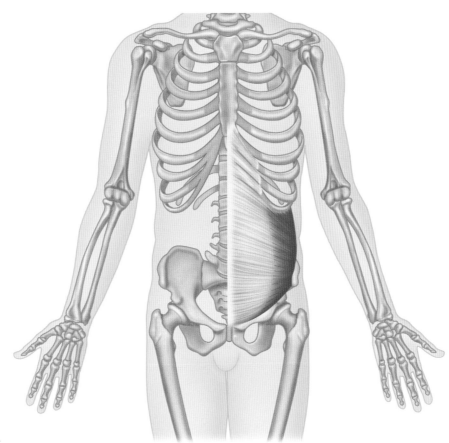

*Anterior view.*

**Latin**, *obliquus*, diagonal, slanted; *internus*, internal; *abdominis*, of the belly/stomach.

### Origin
Iliac crest. Lateral two-thirds of inguinal ligament. Thoracolumbar fascia.

### Insertion
Inferior borders of bottom three or four ribs. Linea alba via an abdominal aponeurosis. Crest of pubis (along with transversus abdominis).

### Action
Compresses abdomen, helping to support abdominal viscera against pull of gravity. Contraction of one side alone laterally bends and rotates trunk.

### Nerve
Ventral rami of thoracic nerves T7–12, ilioinguinal and iliohypogastric nerves.

### Basic functional movement
Example: raking.

### Sports that heavily utilize this muscle
Examples: golf, javelin, pole vault.

### Common problems when muscle is weak
Injury to the lumbar spine, because abdominal muscle tone contributes to stability of this region.

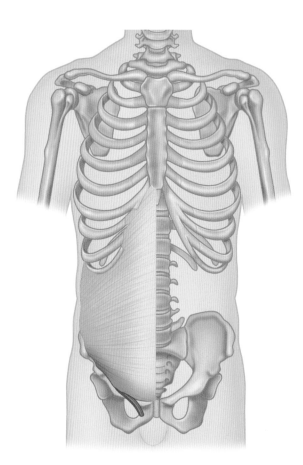

*Anterior view.*

**Greek**, *kremaster*, suspender.

In males, the cremaster is usually well developed; in females, it is underdeveloped or absent. Cremaster forms a thin network of muscle fibers around the spermatic cord and testes (or around the distal portion of the round ligament of the uterus).

**Origin**
Inguinal ligament.

**Insertion**
Pubic tubercle. Crest of pubis. Sheath of rectus abdominis.

**Action**
Pulls testes up from scrotum toward body (mainly to regulate temperature of testes).

**Nerve**
Genital branch of genitofemoral nerve, L1, 2.

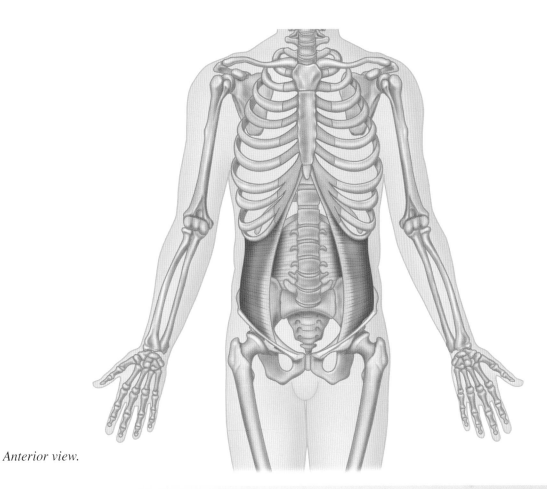

*Anterior view.*

**Latin**, *transversus*, across, crosswise; *abdominis*, of the belly/stomach.

### Origin
Anterior two-thirds of iliac crest. Lateral third of inguinal ligament. Thoracolumbar fascia. Costal cartilages of lower six ribs. Fascia covering iliopsoas.

### Insertion
Xiphoid process and linea alba via an abdominal aponeurosis, the lower fibers of which ultimately attach to the pubic crest and pecten pubis via the conjoint tendon.

### Action
Compresses abdomen, helping to support abdominal viscera against pull of gravity.

### Nerve
Ventral rami of thoracic nerves T7–12, ilioinguinal and iliohypogastric nerves.

### Basic functional movement
Important during forced expiration, sneezing, and coughing. Helps maintain good posture.

### Sports that heavily utilize this muscle
Examples: gymnastics, seated rowing, javelin, pole vault.

### Common problems when muscle is weak
Injury to the lumbar spine, because abdominal muscle tone contributes to stability of this region.

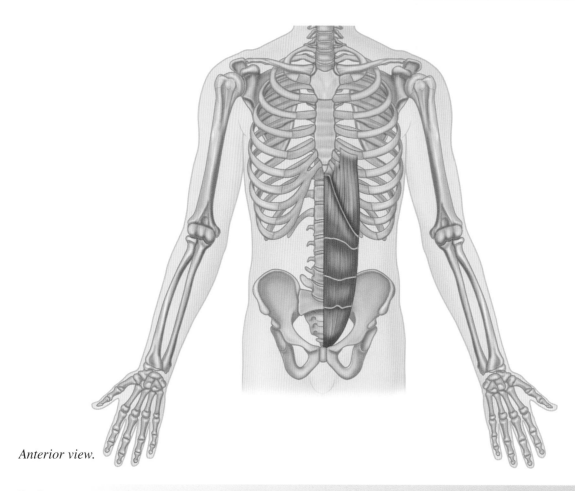

*Anterior view.*

**Latin**, *rectus*, straight; *abdominis*, of the belly/stomach.

Rectus abdominis consists of tendinous bands divided into three or four bellies, each sheathed in aponeurotic fibers from the lateral abdominal muscles. These fibers converge centrally to form the linea alba. Situated anterior to the lower part of rectus abdominis is a frequently absent muscle called *pyramidalis*, which arises from the pubic crest and inserts into the linea alba. It tenses the linea alba, for reasons unknown.

**Origin**
Pubic crest and symphysis pubis.

**Insertion**
Anterior surface of xiphoid process. Fifth, sixth, and seventh costal cartilages.

**Action**
Flexes lumbar spine. Depresses ribcage. Stabilizes pelvis during walking.

**Nerve**
Ventral rami of thoracic nerves T5–12.

**Basic functional movement**
Example: initiating getting out of a low chair.

**Sports that heavily utilize this muscle**
All sports.

**Common problems when muscle is weak**
Injury to the lumbar spine, because abdominal muscle tone contributes to stability of this region.

# Muscles of the Posterior Abdominal Wall

The posterior abdominal wall comprises quadratus lumborum, with the origin of psoas major positioned medial to it, covering the sides of the lumbar vertebral bodies and the anterior aspects of their transverse processes. Psoas major runs downward, to be joined by iliacus, which lines the iliac fossa. Together, these muscles act as padding for various abdominal viscera, and leave the abdomen to become the main flexor of the hip joint.

**Strengthen**

*Dumbbell walking lunge*       *Hanging knee raise*       *Lunges*

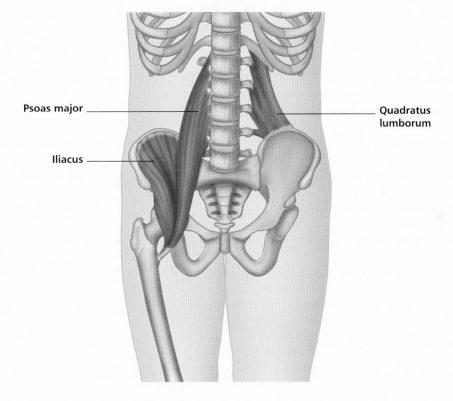

Psoas major

Iliacus

Quadratus lumborum

**Self-stretch**

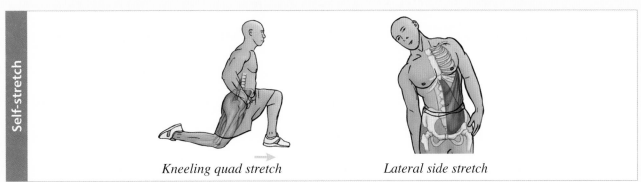

*Kneeling quad stretch*       *Lateral side stretch*

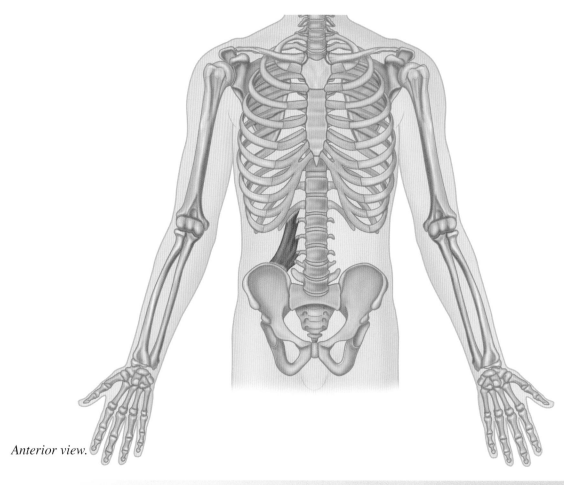

*Anterior view.*

**Latin**, *quadratus*, squared; *lumborum*, of the loins.

**Origin**
Posterior part of iliac crest. Iliolumbar ligament.

**Insertion**
Medial part of lower border of twelfth rib. Transverse processes of upper four lumbar vertebrae (L1–4).

**Action**
Laterally flexes vertebral column. Fixes twelfth rib during deep respiration (e.g. helps stabilize diaphragm for singers exercising voice control). Helps extend lumbar part of vertebral column and gives it lateral stability.

**Nerve**
Ventral rami of subcostal nerve and upper three or four lumbar nerves T12, L**1**, **2**, **3**.

**Basic functional movement**
Example: bending sideways from sitting to pick up an object from the floor.

**Sports that heavily utilize this muscle**
Examples: gymnastics (pommel horse), javelin, tennis serve.

**Movements or injuries that may damage this muscle**
Bending sideways or lifting from a sideways position too quickly.

**Common problems when muscle is chronically tight/shortened (spastic)**
Referred pain to the hip and gluteal area, as well as the low back.

# PSOAS MAJOR (Part of Iliopsoas)

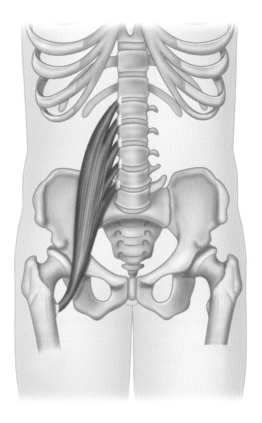

*Anterior view.*

**Greek**, *psoa*, muscle of the loin.
**Latin**, *major*, larger.

Psoas major and iliacus are considered part of the posterior abdominal wall because of their position and their role of cushioning for the abdominal viscera. However, on the basis of their action of flexing the hip joint, it would also be relevant to place these two muscles in Chapter 8, "Muscles of the Hip and Thigh". Note that some of the upper fibers of psoas major may insert by a long tendon into the iliopubic eminence to form psoas minor, which has little function and is absent in about forty percent of people.

Bilateral contraction of psoas major will increase lumbar lordosis.

### Origin
Bases of transverse processes of all lumbar vertebrae (L1–5). Bodies of twelfth thoracic and all lumbar vertebrae (T12–L5). Intervertebral discs above each lumbar vertebra.

### Insertion
Lesser trochanter of femur.

### Action
Main flexor of hip joint, in conjunction with iliacus (flexes and laterally rotates thigh, as in kicking a football). Acting from its insertion, it flexes trunk, as in sitting up from supine position.

### Nerve
Ventral rami of lumbar nerves L1, **2**, **3**, 4 (psoas minor innervated from L1, **2**).

### Basic functional movement
Example: going up a step or walking up an incline.

### Sports that heavily utilize this muscle
Examples: rock-face climbing, sprinting (maximizes stride length), kicking sports (e.g. soccer, to maximize kicking force).

### Common problems when muscle is chronically tight/shortened (spastic)
Low back pain due to an increase in lumbar curve (lordosis).

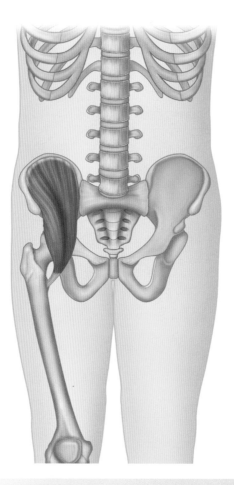

*Anterior view.*

**Latin**, *iliacus*, relating to the loin.

**Origin**
Superior two-thirds of iliac fossa. Internal lip of iliac crest. Ala of sacrum and anterior ligaments of lumbosacral and sacroiliac joints.

**Insertion**
Lateral side of tendon of psoas major, continuing into lesser trochanter of femur.

**Action**
Main flexor of hip joint, in conjunction with psoas major. Flexes and laterally rotates thigh, as in kicking a football. Brings leg forward in walking or running. Acting from its insertion, it flexes the trunk, as in sitting up from supine position.

**Nerve**
Femoral nerve L(1), **2**, **3**, 4.

**Basic functional movement**
Example: going up a step or walking up an incline.

**Sports that heavily utilize this muscle**
Examples: rock-face climbing, sprinting (maximizes stride length), kicking sports (e.g. soccer, to maximize kicking force).

**Common problems when muscle is chronically tight/shortened (spastic)**
Low back pain due to an increase in lumbar curve (lordosis).

# Muscles of the Shoulder and Arm

The upper limb has evolved to provide us with the specialty of manipulation and dexterity, whereas the lower limb provides locomotion. This being the case, the upper limb emphasizes mobility while sacrificing stability. Mobility of the upper limb is mainly dependent on three joints: sternoclavicular, acromioclavicular, and glenohumeral. Muscles in this area can be categorized according to: 1) muscles that run between the trunk and the scapula, which act upon the shoulder girdle and not the shoulder joint, i.e. **trapezius**, **levator scapulae**, **rhomboids**, **serratus anterior**, **pectoralis minor**, and **subclavius**; 2) muscles that run between the trunk and the humerus, which act upon the shoulder joint and the shoulder girdle, i.e. **pectoralis major** and **latissimus dorsi**; and 3) muscles that run between the scapula and the humerus, which act exclusively upon the shoulder joint, i.e. **deltoideus**, **supraspinatus**, **infraspinatus**, **teres minor**, **subscapularis**, **teres major**, and **coracobrachialis**.

**Latissimus dorsi**, the broadest muscle of the back, is one of the chief climbing muscles, since it pulls the shoulders downward and backward, and pulls the trunk up to the fixed arms. It is therefore heavily used in sports such as climbing, gymnastics (in particular rings and parallel bars), swimming, and rowing. The **rhomboid** muscles are situated between the scapula and the vertebral column, and are so named because of their shape (rhombus), with rhomboideus major being larger than rhomboideus minor.

It is recommended that any exercises for the rotator cuff muscles—**subscapularis**, **infraspinatus**, **supraspinatus**, and **teres minor**—should be performed primarily from the standing position if possible. The rotator cuff muscles are less concerned with strength and more concerned with dexterity (wider range of motion), which means stability is compromised. Forces generated in the lower limbs translate via the thorax to the glenohumeral complex. Sitting while performing shoulder movements such as internal or external rotations with flexion or extension leave the shoulder hanging out on its own. Standing positions are more functional, whereas the tasks required when seated should be simple.

The muscles of the arm comprise those that originate from the scapula and/or humerus, and insert into the radius and/or ulna; they therefore act upon the elbow joint and/or superior radioulnar joint. The muscles in question are biceps brachii, brachialis, triceps brachii, and anconeus. **Coracobrachialis**, although acting upon the shoulder joint, is also included because of its proximity to the other muscles of this group. **Biceps brachii** operates over three joints, and has two tendinous heads at its origin and two tendinous insertions. Occasionally it has a third head, originating at the insertion of coracobrachialis. The short head forms part of the lateral wall of the axilla, along with coracobrachialis and the humerus. **Brachialis** lies posterior to biceps brachii and is the main flexor of the elbow joint. **Triceps brachii**, which originates from three heads, and **anconeus** are the only muscles on the posterior arm.

# Muscles Attaching the Upper Limb to the Trunk

This section includes the group of muscles that run between the trunk and the scapula, which act upon the shoulder girdle and not the shoulder joint, and those muscles that run between the trunk and the humerus, which act upon the shoulder joint and the shoulder girdle.

Alternate dumbbell curl        Flat dumbbell flye        Standing dumbbell press

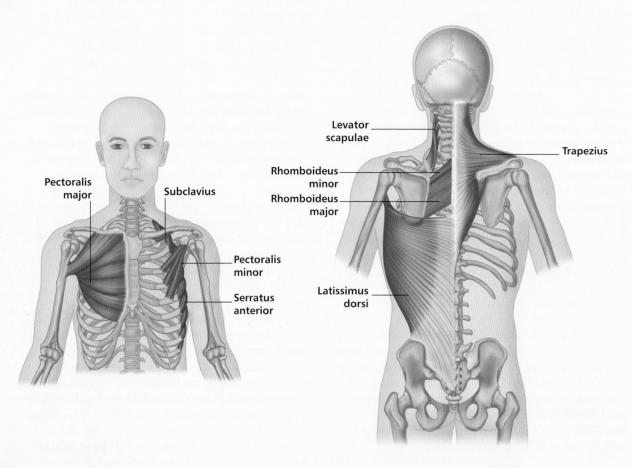

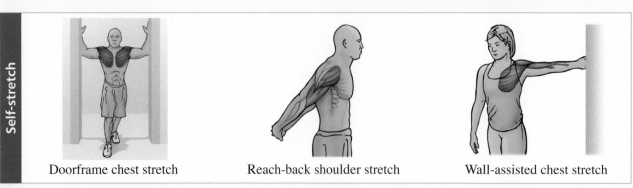

Doorframe chest stretch        Reach-back shoulder stretch        Wall-assisted chest stretch

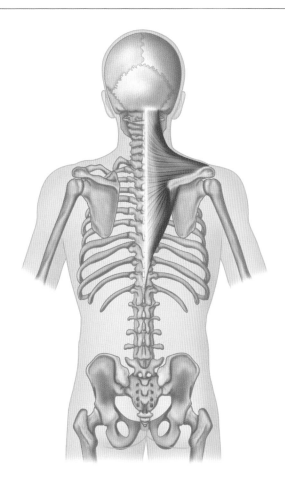

**Greek**, *trapezoeides*, table shaped.

The left and right trapezius viewed as a whole create a trapezium in shape, thus giving this muscle its name.

### Origin
Medial third of superior nuchal line of occipital bone. External occipital protuberance. Ligamentum nuchae. Spinous processes and supraspinous ligaments of seventh cervical vertebra (C7) and all thoracic vertebrae (T1–12).

### Insertion
Posterior border of lateral third of clavicle. Medial border of acromion. Upper border of crest of spine of scapula, and tubercle on this crest.

### Action
Upper fibers: pull shoulder girdle up (elevation). Help prevent depression of shoulder girdle when a weight is carried on the shoulder or in the hand.
Middle fibers: retract (adduct) scapula.
Lower fibers: depress scapula, particularly against resistance, as when using hands to get up from a chair.
Upper and lower fibers together: rotate scapula, as in elevating the arm above the head.

### Nerve
Motor supply: accessory **XI** nerve.
Sensory supply (proprioception): ventral ramus of cervical nerves C2, **3**, **4**.

### Basic functional movement
Example (upper and lower fibers working together): painting a ceiling.

### Sports that heavily utilize this muscle
Examples: shot put, boxing, seated rowing.

### Common problems when muscle is chronically tight/ shortened (spastic)
Upper fibers: neck pain or stiffness, headaches.

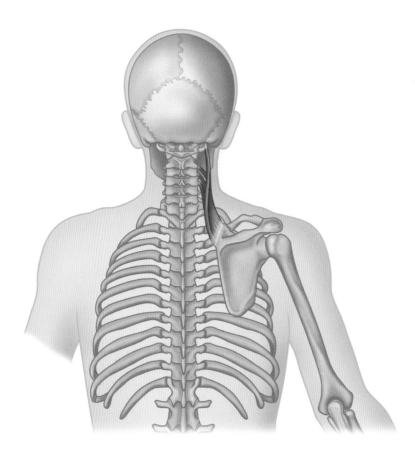

**Latin**, *levare*, to lift; *scapulae*, of the shoulder blade.

Levator scapulae is deep to sternocleidomastoideus and trapezius. It is named after its action of elevating the scapula.

**Origin**
Posterior tubercles of transverse processes of first three or four cervical vertebrae (C1–4).

**Insertion**
Medial (vertebral) border of scapula, between superior angle and spine of scapula.

**Action**
Elevates scapula. Helps retract scapula. Helps bend neck laterally.

**Nerve**
Dorsal scapular nerve C**4**, **5**, and cervical nerves C**3**, **4**.

**Basic functional movement**
Example: carrying a heavy bag.

**Sports that heavily utilize this muscle**
Examples: shot put, weightlifting.

**Common problems when muscle is chronically tight/ shortened (spastic)**
Upper fibers: neck pain or stiffness, headaches.

# RHOMBOIDEUS MINOR

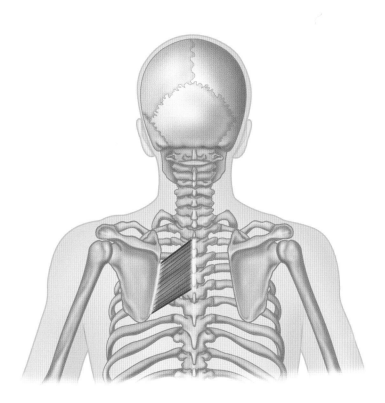

**Greek**, *rhomboeides*, parallelogram shaped, with only opposite sides and angles equal. **Latin**, *minor*, smaller.

Rhomboideus minor connects the scapula with the vertebrae, and lies deep to trapezius. So named because of its shape.

**Origin**
Spinous processes and supraspinous ligaments of seventh cervical and first thoracic vertebrae. Lower part of ligamentum nuchae.

**Insertion**
Medial (vertebral) border of scapula at the level of spine of scapula.

**Action**
Retracts (adducts) scapula. Stabilizes scapula. Slightly elevates medial border of scapula, causing downward rotation (therefore depressing the lateral angle). Slightly assists in outer range of adduction of arm (i.e. from arm overhead to arm at shoulder level).

**Nerve**
Dorsal scapular nerve C4, 5.

**Basic functional movement**
Example: pulling something toward you, such as opening a drawer.

**Sports that heavily utilize this muscle**
Examples: archery, seated rowing, windsurfing, racket sports.

**Common problems when muscle is tight or overstretched**
Tight: soreness or aching between the shoulder blades.
Overstretched: rounded shoulders are both symptomatic of, and exacerbated by, overstretched rhomboids (which tend to get overstretched rather than becoming too tight).

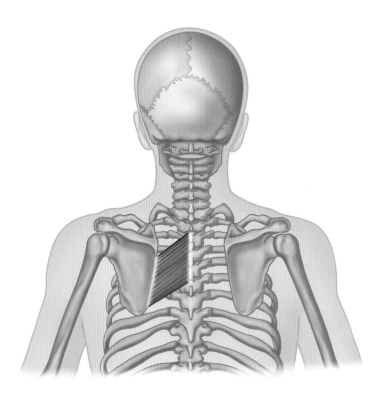

**Greek**, *rhomboeides*, parallelogram shaped, with only opposite sides and angles equal. **Latin**, *major*, larger.

Rhomboideus major runs parallel to, and is often continuous with, rhomboideus minor. It is so named because of its shape.

**Origin**
Spinous processes and supraspinous ligaments of second to fifth thoracic vertebrae (T2–5).

**Insertion**
Medial border of scapula, between spine of scapula and inferior angle.

**Action**
Retracts (adducts) scapula. Stabilizes scapula. Slightly elevates medial border of scapula, causing downward rotation. Slightly assists in outer range of adduction of arm (i.e. from arm overhead to arm at shoulder level).

**Nerve**
Dorsal scapular nerve C4, 5.

**Basic functional movement**
Example: pulling something toward you, such as opening a drawer.

**Sports that heavily utilize this muscle**
Examples: archery, seated rowing, windsurfing, racket sports.

**Common problems when muscle is tight or overstretched**
Tight: soreness or aching between the shoulder blades.
Overstretched: rounded shoulders are both symptomatic of, and exacerbated by, overstretched rhomboids (which tend to get overstretched rather than becoming too tight).

# SERRATUS ANTERIOR

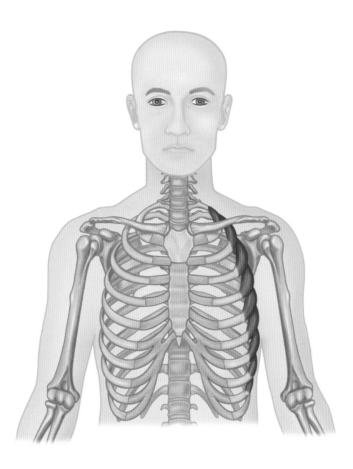

**Latin**, *serratus*, serrated; *anterior*, at the front.

Serratus anterior forms the medial wall of the axilla, along with the upper five ribs. It is a large muscle composed of a series of finger-like slips. The lower slips interdigitate with the origin of the external oblique.

### Origin
Outer surfaces and superior borders of upper eight or nine ribs, and fascia covering their intercostal spaces.

### Insertion
Anterior (costal) surface of medial border of scapula and inferior angle of scapula.

### Action
Rotates scapula for abduction and flexion of arm. Protracts scapula (pulls it forward on the chest wall and holds it closely in to the chest wall), facilitating pushing movements, such as press-ups or punching.

### Nerve
Long thoracic nerve C**5**, **6**, **7**, 8.

### Basic functional movement
Example: reaching forward for something barely within reach.

### Sports that heavily utilize this muscle
Examples: boxing, shot put.

### Common problems when muscle is weak
Note: A lesion of the long thoracic nerve will result in the medial border of the scapula falling away from the posterior chest wall, resulting in a "winged scapula" (looking like an angel's wing). This is also a feature when the nerve to this muscle is damaged.

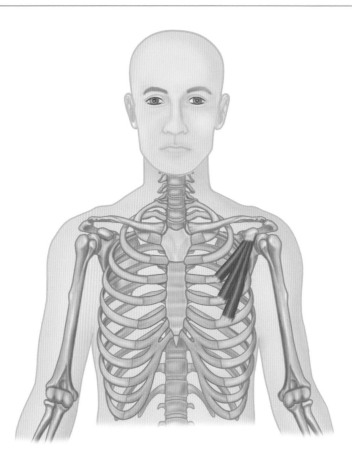

**Latin**, *pectoralis*, relating to the chest; *minor*, smaller.

Pectoralis minor is a flat triangular muscle lying posterior to, and concealed by, pectoralis major. Along with pectoralis major, it forms the anterior wall of the axilla.

**Origin**
Outer surfaces of third, fourth, and fifth ribs, and fascia of the corresponding intercostal spaces.

**Insertion**
Coracoid process of scapula.

**Action**
Draws scapula forward and downward. Raises ribs during forced inspiration (i.e. it is an accessory muscle of inspiration, if the scapula is stabilized by the rhomboids and trapezius).

**Nerve**
Medial pectoral nerve, with fibers from a communicating branch of lateral pectoral nerve, C(6), **7**, **8**, T1.

**Basic functional movement**
Example: pushing on the arms of a chair to stand up.

**Sports that heavily utilize this muscle**
Racket sports, e.g. tennis, badminton. Baseball pitching. Sprinting.

**Common problems when muscle is chronically tight/ shortened (spastic)**
Restricts expansion of the chest.

# SUBCLAVIUS

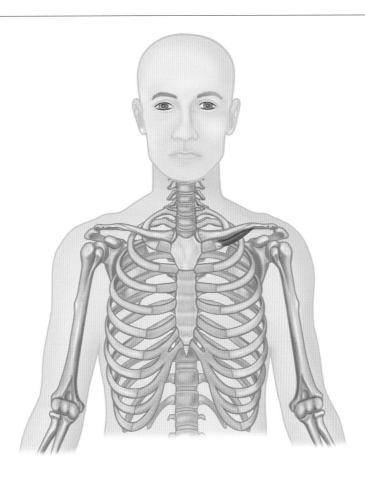

**Latin**, *sub*, under; *clavis*, key.

Subclavius is posterior to, and concealed by, the clavicle and pectoralis major. Paralysis of this muscle produces no apparent effect.

### Origin
Junction of first rib and first costal cartilage.

### Insertion
Floor of a groove on lower (inferior) surface of clavicle.

### Action
Depresses clavicle and draws it toward sternum, thereby steadying it in movements of shoulder girdle.

### Nerve
Nerve to subclavius C**5**, **6**.

# PECTORALIS MAJOR

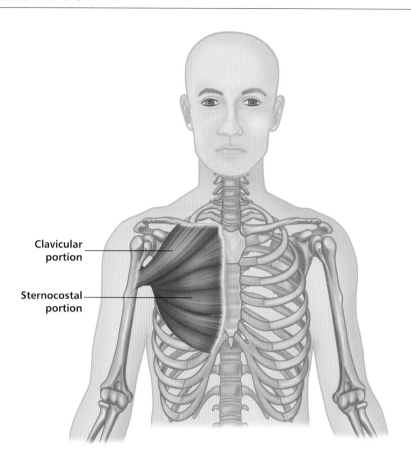

Clavicular portion

Sternocostal portion

**Latin**, *pectoralis*, relating to the chest; *major*, larger.

Along with pectoralis minor, pectoralis major forms the anterior wall of the axilla.

## Origin
Clavicular head: medial half or two-thirds of front of clavicle. Sternocostal portion: front of manubrium and body of sternum. Upper six costal cartilages. Rectus sheath.

## Insertion
Crest below greater tubercle of humerus. Lateral lip of intertubercular sulcus (bicipital groove) of humerus.

## Action
Adducts and medially rotates humerus.
Clavicular portion: flexes and medially rotates shoulder joint, and horizontally adducts humerus toward opposite shoulder.
Sternocostal portion: obliquely adducts humerus toward opposite hip.
Pectoralis major is one of the main climbing muscles, pulling the body up to the fixed arm.

## Nerve
Nerve to upper fibers: lateral pectoral nerve C**5**, **6**, **7**.
Nerve to lower fibers: lateral and medial pectoral nerves C**6**, **7**, **8**, T**1**.

## Basic functional movement
Clavicular portion: brings the arm forward and across the body, e.g. as in applying deodorant to the opposite armpit.
Sternocostal portion: pulling something down from above, e.g. a rope in bell-ringing.

## Sports that heavily utilize this muscle
Examples: racket sports (e.g. tennis), golf, baseball pitching, gymnastics (rings and high bar), judo, wrestling.

## Movements or injuries that may damage this muscle
Indian wrestling and other strength activities that force medial rotation and adduction can damage the insertion of this muscle.

## Common problems when muscle is tight
Rounds the back and restricts expansion of the chest, restricting lateral rotation and abduction of the shoulder.

# LATISSIMUS DORSI

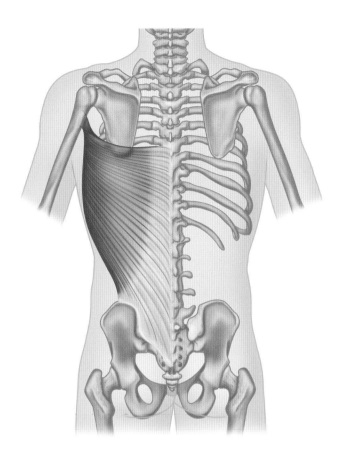

*Posterior view.*

**Latin**, *latissimus*, widest; *dorsi*, of the back.

Along with subscapularis and teres major, latissimus dorsi forms the posterior wall of the axilla.

## Origin
Thoracolumbar fascia, which is attached to spinous processes of lower six thoracic vertebrae and all lumbar and sacral vertebrae (T7–S5) and to intervening supraspinous ligaments. Posterior part of iliac crest. Lower three or four ribs. Inferior angle of scapula.

## Insertion
Floor of intertubercular sulcus (bicipital groove) of humerus.

## Action
Extends flexed arm. Adducts and medially rotates humerus. It is one of the chief climbing muscles, since it pulls shoulders downward and backward, and pulls trunk up to the fixed arms (therefore also active in crawl swimming stroke). Assists in forced inspiration, by raising lower ribs.

## Nerve
Thoracodorsal nerve C**6**, **7**, **8**, from the posterior cord of brachial plexus.

## Basic functional movement
Example: pushing on the arms of a chair to stand up.

## Sports that heavily utilize this muscle
Examples: climbing, gymnastics (rings, parallel bars), swimming, rowing.

# Muscles of the Shoulder Joint

This section includes the group of muscles that run between the scapula and the humerus, which act exclusively upon the shoulder joint. Coracobrachialis also acts exclusively upon the shoulder joint, but, because of its position, has been included in the muscles of the arm section.

Close-grip bench press

Dumbbell shoulder press

Lateral raise

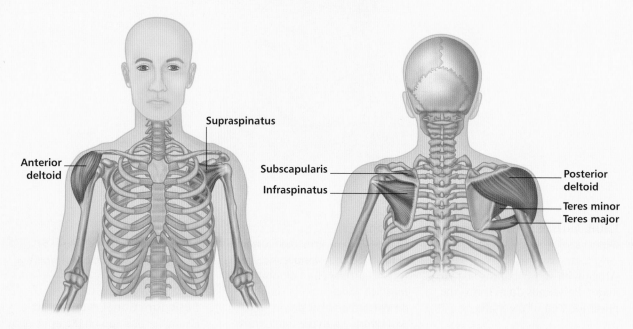

Supraspinatus

Anterior deltoid

Subscapularis

Infraspinatus

Posterior deltoid

Teres minor

Teres major

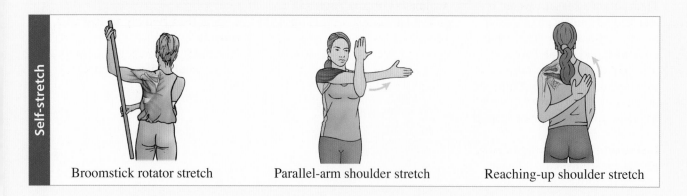

Broomstick rotator stretch

Parallel-arm shoulder stretch

Reaching-up shoulder stretch

# DELTOIDEUS

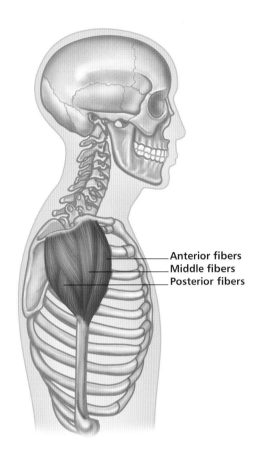

Anterior fibers
Middle fibers
Posterior fibers

**Greek**, *deltoeides*, shaped like the Greek capital letter delta (Δ).

Deltoideus is composed of three parts: anterior, middle, and posterior. Only the middle part is multipennate, probably because its mechanical disadvantage of abduction of the shoulder joint requires extra strength.

## Origin
Anterior fibers: anterior border and superior surface of lateral third of clavicle.
Middle fibers: lateral border of acromion process.
Posterior fibers: lower lip of crest of spine of scapula.

## Insertion
Deltoid tuberosity, situated halfway down lateral surface of shaft of humerus.

## Action
Anterior fibers: flex and medially rotate humerus.
Middle fibers: abduct humerus at shoulder joint (only after the movement has been initiated by supraspinatus).
Posterior fibers: extend and laterally rotate humerus.

## Nerve
Axillary nerve C**5**, **6**, from the posterior cord of brachial plexus.

## Basic functional movement
Examples: reaching for something out to the side, raising the arm to wave.

## Sports that heavily utilize this muscle
Examples: javelin, shot put, racket sports, windsurfing, weightlifting.

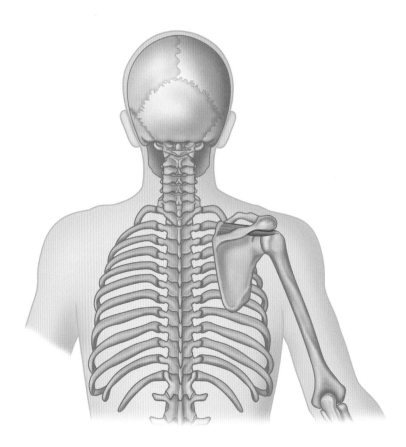

**Latin**, *supra*, above; *spina*, spine.

Supraspinatus is a member of the rotator cuff, which also includes infraspinatus, teres minor, and subscapularis. The rotator cuff helps hold the head of the humerus in contact with the glenoid cavity (fossa, socket) of the scapula during movements of the shoulder, thus helping to prevent dislocation of the joint.

### Origin
Supraspinous fossa of scapula.

### Insertion
Upper aspect of greater tubercle of humerus. Capsule of shoulder joint.

### Action
Initiates the process of abduction at shoulder joint, so that deltoideus can take over in the later stages of abduction.

### Nerve
Suprascapular nerve C4, **5**, 6, from upper trunk of brachial plexus.

### Basic functional movement
Example: holding a shopping bag away from the side of the body.

### Sports that heavily utilize this muscle
Examples: baseball, golf, racket sports.

### Movements or injuries that may damage this muscle
Dislocation of the shoulder joint.

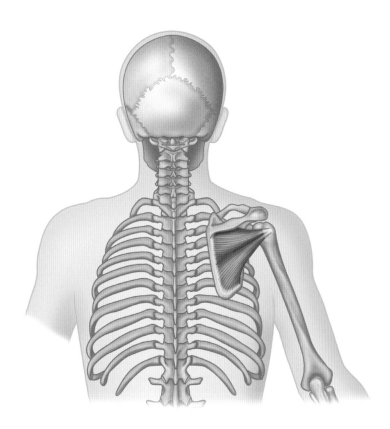

**Latin**, *infra*, below; *spina*, spine.

Infraspinatus is a member of the rotator cuff, which also includes supraspinatus, teres minor, and subscapularis. The rotator cuff helps hold the head of the humerus in contact with the glenoid cavity (fossa, socket) of the scapula during movements of the shoulder, thus helping to prevent dislocation of the joint.

**Origin**
Infraspinous fossa of scapula.

**Insertion**
Middle facet on greater tubercle of humerus. Capsule of shoulder joint.

**Action**
As a rotator cuff muscle, infraspinatus helps prevent posterior dislocation of shoulder joint. Laterally rotates humerus.

**Nerve**
Suprascapular nerve C(4), **5**, **6**, from upper trunk of brachial plexus.

**Basic functional movement**
Example: brushing back hair.

**Sports that heavily utilize this muscle**
Example: backhand racket sports.

**Movements or injuries that may damage this muscle**
Dislocation of the shoulder joint.

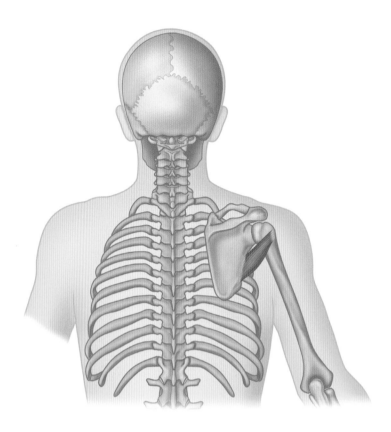

**Latin**, *teres*, rounded, finely shaped; *minor*, smaller.

Teres minor is a member of the rotator cuff, which also includes supraspinatus, infraspinatus, and subscapularis. The rotator cuff helps hold the head of the humerus in contact with the glenoid cavity (fossa, socket) of the scapula during movements of the shoulder, thus helping to prevent dislocation of the joint.

**Origin**
Upper two-thirds of lateral border of dorsal surface of scapula.

**Insertion**
Lower facet on greater tubercle of humerus. Capsule of shoulder joint.

**Action**
As a rotator cuff muscle, teres minor helps prevent upward dislocation of shoulder joint. Laterally rotates humerus. Weakly adducts humerus.

**Nerve**
Axillary nerve C**5**, **6**, from posterior cord of brachial plexus.

**Basic functional movement**
Example: brushing back hair.

**Sports that heavily utilize this muscle**
Example: backhand racket sports.

**Movements or injuries that may damage this muscle**
Dislocation of the shoulder joint.

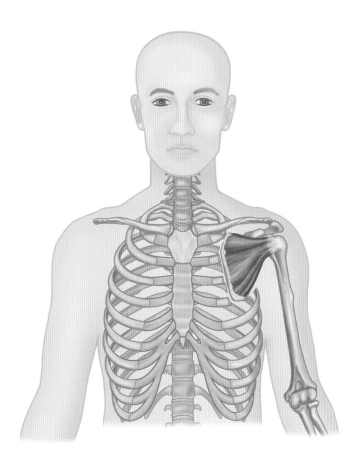

**Latin**, *sub*, under; *scapularis*, relating to the shoulder blade.

Subscapularis is a member of the rotator cuff, which also includes supraspinatus, infraspinatus, and teres minor. The rotator cuff helps hold the head of the humerus in contact with the glenoid cavity (fossa, socket) of the scapula during movements of the shoulder, thus helping to prevent dislocation of the joint. Subscapularis constitutes the greater part of the posterior wall of the axilla.

**Origin**
Subscapular fossa and groove along lateral border of anterior surface of scapula.

**Insertion**
Lesser tubercle of humerus. Capsule of shoulder joint.

**Action**
As a rotator cuff muscle, subscapularis stabilizes the glenohumeral joint, mainly preventing the head of the humerus being pulled upward by deltoideus, biceps brachii, and long head of triceps brachii. Medially rotates humerus.

**Nerve**
Upper and lower subscapular nerves C**5**, **6**, 7, from posterior cord of brachial plexus.

**Basic functional movement**
Example: reaching into the back pocket.

**Sports that heavily utilize this muscle**
Examples: athletic throwing events, golf, racket sports.

**Movements or injuries that may damage this muscle**
Twisting the arm behind the back (as in an overzealous restraining hold), or struggling to free oneself from that position, may damage the insertion.

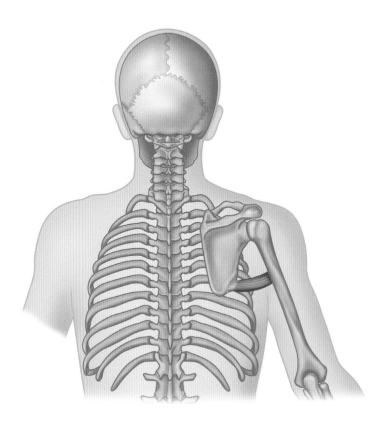

**Latin**, *teres*, rounded, finely shaped; *major*, larger.

Teres major, along with the tendon of latissimus dorsi (which passes around it) and the subscapularis, forms the posterior fold of the axilla.

**Origin**
Oval area on lower third of posterior surface of lateral border of scapula.

**Insertion**
Medial lip of intertubercular sulcus (bicipital groove) of humerus.

**Action**
Adducts humerus. Medially rotates humerus. Extends humerus from flexed position.

**Nerve**
Lower subscapular nerve C5, **6**, 7, from posterior cord of brachial plexus.

**Basic functional movement**
Example: reaching into the back pocket.

**Sports that heavily utilize this muscle**
Examples: rowing, cross-country skiing.

**Movements or injuries that may damage this muscle**
Jerking the arm sharply forward, as in throwing a stone to skim it across a lake.

# Muscles of the Arm

The muscles of the arm comprise those that originate from the scapula and/or humerus, and insert into the radius and/or ulna; they therefore act upon the elbow joint. Coracobrachialis, although acting upon the shoulder joint, is also included here because of its proximity to the other muscles of this group.

**Strengthen**

Dumbbell kickbacks         Dumbbell curls         Rope press-down

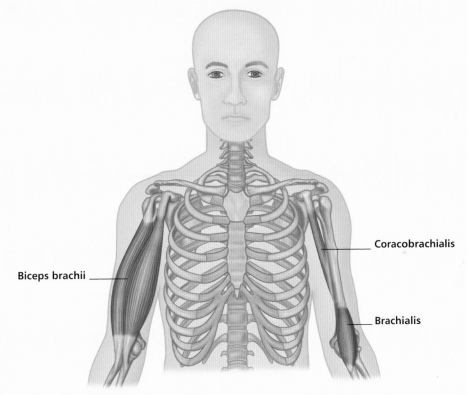

Coracobrachialis

Biceps brachii

Brachialis

**Self-stretch**

Assisted reverse shoulder stretch     Kneeling forearm stretch     Overhead triceps stretch

# BICEPS BRACHII

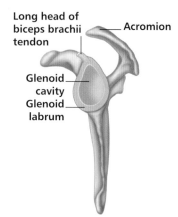

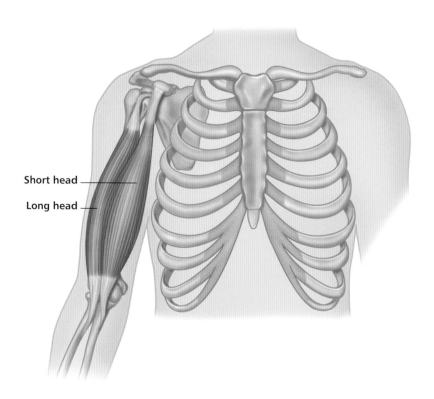

Long head of biceps brachii tendon

Acromion

Glenoid cavity

Glenoid labrum

Short head

Long head

*Anterior view.*

**Latin**, *biceps*, two-headed; *brachii*, of the arm.

Biceps brachii operates over three joints. It has two tendinous heads at its origin and two tendinous insertions; occasionally it has a third head, originating at the insertion of coracobrachialis. The short head forms part of the lateral wall of the axilla, along with coracobrachialis and the humerus.

**Origin**
Short head: tip of coracoid process of scapula.
Long head: supraglenoid tubercle of scapula.

**Insertion**
Posterior part of radial tuberosity. Bicipital aponeurosis, which leads into deep fascia on medial aspect of forearm.

**Action**
Flexes elbow joint. Supinates forearm. (It has been described as the muscle that puts in the corkscrew and pulls out the cork). Weakly flexes arm at shoulder joint.

**Nerve**
Musculocutaneous nerve C5, 6.

**Basic functional movement**
Examples: picking up an object, bringing food to the mouth.

**Sports that heavily utilize this muscle**
Examples: boxing, climbing, canoeing, rowing.

**Movements or injuries that may damage this muscle**
Lifting heavy objects too quickly.

**Common problems when muscle is chronically tight/ shortened**
Flexion deformity of the elbow (elbow cannot be fully straightened).

# CORACOBRACHIALIS

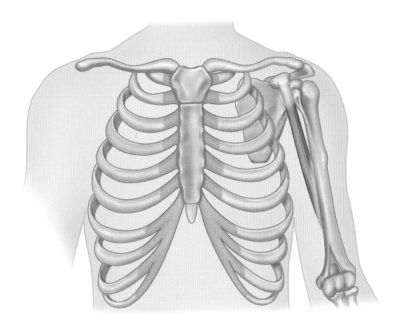

*Anterior view.*

**Greek**, *korakoeides*, raven-like.
**Latin**, *brachialis*, relating to the arm.

Along with the short head of biceps brachii and the humerus, coracobrachialis forms the lateral wall of the axilla. Coracobrachialis is so named because it resembles a raven's beak.

**Origin**
Tip of coracoid process of scapula.

**Insertion**
Medial aspect of humerus at mid-shaft.

**Action**
Weakly adducts shoulder joint. Possibly assists in flexion of shoulder joint (but this has not been proved). Helps stabilize humerus.

**Nerve**
Musculocutaneous nerve C**6**, **7**.

**Basic functional movement**
Example: mopping the floor.

**Sports that heavily utilize this muscle**
Examples: golf, cricket batting.

**Movements or injuries that may damage this muscle**
Suddenly hitting the ground when swinging the bat hard in cricket.

# BRACHIALIS

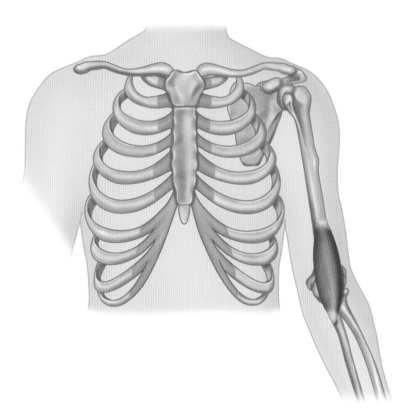

*Anterior view.*

**Latin**, *brachialis*, relating to the arm.

Brachialis lies posterior to biceps brachii and is the main flexor of the elbow joint. Some fibers may be partly fused with brachioradialis.

**Origin**
Lower (distal) two-thirds of anterior aspect of humerus.

**Insertion**
Coronoid process of ulna and tuberosity of ulna (i.e. area on front of upper part of shaft of ulna).

**Action**
Flexes elbow joint.

**Nerve**
Musculocutaneous nerve C**5**, **6**.

**Basic functional movement**
Example: bringing food to the mouth.

**Sports that heavily utilize this muscle**
Examples: baseball, boxing, gymnastics.

**Common problems when muscle is chronically tight/ shortened**
Flexion deformity of the elbow (elbow cannot be fully straightened).

# TRICEPS BRACHII

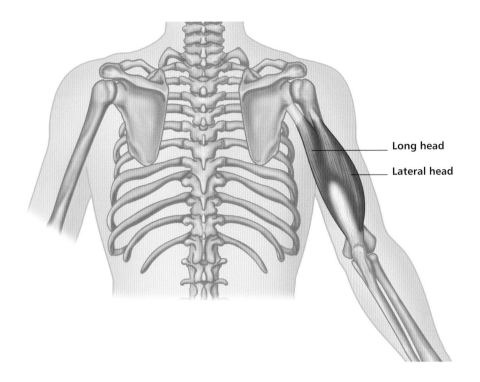

Long head

Lateral head

*Posterior view.*

**Latin**, *triceps*, three-headed; *brachii*, of the arm.

Triceps brachii originates from three heads and is the only muscle on the back of the arm. The medial head is largely covered by the lateral and long heads.

### Origin
Long head: infraglenoid tubercle of scapula.
Lateral head: upper half of posterior surface of shaft of humerus (above and lateral to radial groove).
Medial head: lower half of posterior surface of shaft of humerus (below and medial to radial groove).

### Insertion
Posterior part of olecranon process of ulna.

### Action
Extends elbow joint. Long head can adduct humerus and extend it from flexed position. Stabilizes shoulder joint.

### Nerve
Radial nerve C6, **7**, **8**, T1.

### Basic functional movement
Examples: throwing objects, pushing a door shut.

### Sports that heavily utilize this muscle
Examples: basketball or netball (shooting), shot put, baseball (pitching), volleyball.

### Movements or injuries that may damage this muscle
Throwing with excessive force.

### Problems when muscle is chronically tight/shortened
Extension deformity of elbow (elbow cannot be fully flexed), although not very common.

# ANCONEUS

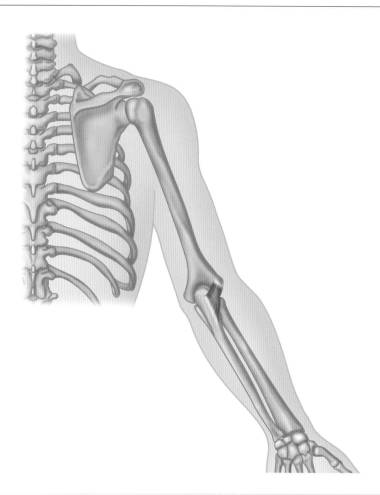

*Posterior view.*

**Greek**, *agkon*, elbow.

**Origin**
Posterior part of lateral epicondyle of humerus.

**Insertion**
Lateral surface of olecranon process and upper portion of posterior surface of ulna.

**Action**
Assists triceps in extending forearm at elbow joint. May stabilize ulna during pronation and supination.

**Nerve**
Radial nerve C**7**, **8**.

**Basic functional movement**
Example: pushing objects at arm's length.

# Muscles of the Forearm and Hand

The anterior forearm contains three functional muscle groups, in a superficial to deep arrangement: the pronators of the forearm, the wrist flexors, and the long flexors of the fingers and thumb. The superficial tissue comprises four muscles: **pronator teres**, **flexor carpi radialis**, **palmaris longus**, and **flexor carpi ulnaris**—all emanating from a common tendon known as the *common flexor origin*. The middle tissues contain only the **flexor digitorum superficialis**. The deepest tissues consist of **flexor digitorum profundus**, **flexor pollicis longus**, and **pronator quadratus**.

On the posterior aspect of the forearm there are two muscle groups—superficial and deep. The superficial group contains, from the radial to the ulnar side, **brachioradialis**, **extensor carpi radialis longus**, **extensor carpi radialis brevis**, **extensor digitorum**, **extensor digiti minimi**, and **extensor carpi ulnaris**. The muscle belly of **brachioradialis** is prominent when working against resistance. The deep group contains **supinator**, **abductor pollicis longus**, **extensor pollicis brevis**, **extensor pollicis longus**, and **extensor indicis**.

The muscle groupings in the hand are: 1) the intrinsic muscles, consisting of **lumbricales**, which arise from the tendons of flexor digitorum profundus in the palm and act on the four fingers, and **palmar** and **dorsal interossei**, located within the intermetacarpal spaces to act on the four fingers and thumb; 2) the muscles of the hypothenar eminence—**abductor digiti minimi**, **opponens digiti minimi**, **flexor digiti minimi brevis**, and **palmaris longus**; 3) the muscles of the thenar eminence—**abductor pollicis brevis**, **opponens pollicis**, and **flexor pollicis brevis**; and 4) **adductor pollicis**.

# Muscles of the Anterior Forearm

The anterior forearm contains three functional muscle groups: the pronators of the forearm, the wrist flexors, and the long flexors of the fingers and thumb. They are arranged in three layers: superficial, middle, and deep.

**Strengthen**

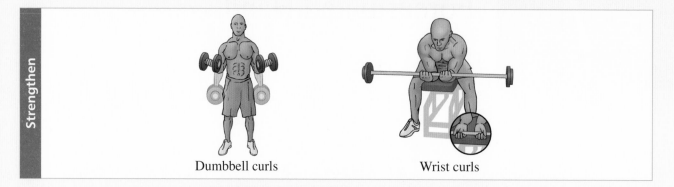

Dumbbell curls

Wrist curls

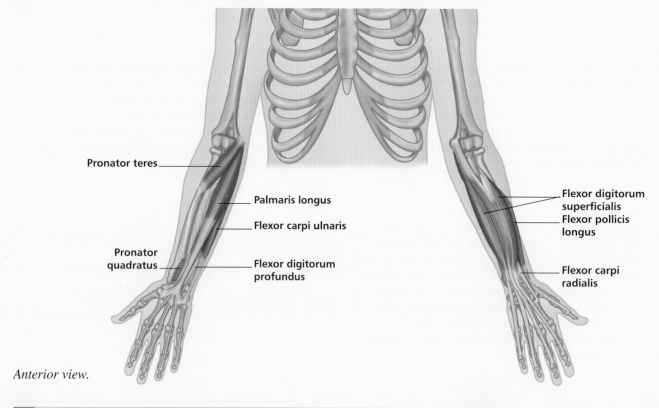

Pronator teres

Palmaris longus

Flexor carpi ulnaris

Pronator quadratus

Flexor digitorum profundus

Flexor digitorum superficialis

Flexor pollicis longus

Flexor carpi radialis

*Anterior view.*

**Self-stretch**

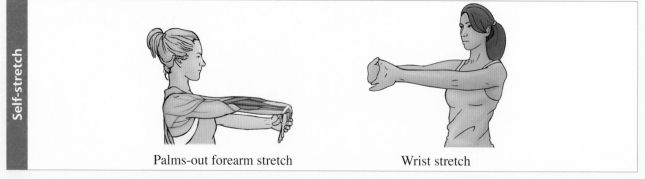

Palms-out forearm stretch

Wrist stretch

# PRONATOR TERES

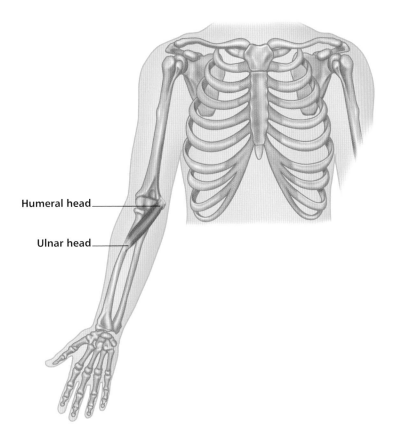

Humeral head

Ulnar head

*Anterior view.*

**Latin**, *pronare*, to bend forward; *teres*, rounded, finely shaped.

Part of the superficial layer of the anterior forearm, which also includes flexor carpi radialis, palmaris longus, and flexor carpi ulnaris.

**Origin**
Humeral head: lower third of medial supracondylar ridge and common flexor origin on anterior aspect of medial epicondyle of humerus.
Ulnar head: medial border of coronoid process of ulna.

**Insertion**
Mid-lateral surface of radius (pronator tuberosity).

**Action**
Pronates forearm. Assists in flexion of elbow joint.

**Nerve**
Median nerve C**6**, **7**.

**Basic functional movement**
Examples: pouring liquid from a container, turning a doorknob.

**Sports that heavily utilize this muscle**
Examples: cricket batting, hockey dribbling, volleyball smash.

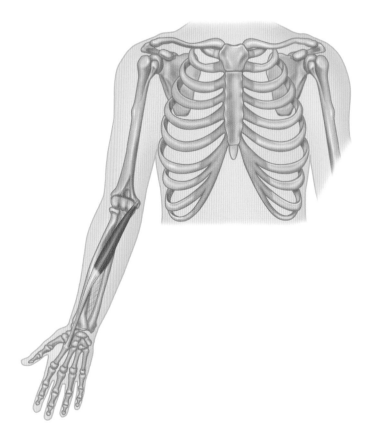

*Anterior view.*

**Latin**, *flectere*, to bend; *carpi*, of the wrist; *radius*, staff, spoke of wheel.

Part of the superficial layer of the anterior forearm, which also includes: pronator teres, palmaris longus, and flexor carpi ulnaris.

### Origin
Common flexor origin on anterior aspect of medial epicondyle of humerus.

### Insertion
Front of bases of second and third metacarpal bones.

### Action
Flexes and abducts carpus (wrist joint). Helps to flex elbow and pronate forearm.

### Nerve
Median nerve C**6**, **7**, 8.

### Basic functional movement
Examples: pulling rope in toward you, wielding an axe or hammer.

### Sports that heavily utilize this muscle
Examples: sailing, waterskiing, golf, baseball, cricket, volleyball.

### Movements or injuries that may damage this muscle
Over-extending the wrist as a result of breaking a fall with the hand.

### Common problems when muscle is chronically tight/ shortened (spastic)
Golfer's elbow (overuse tendonitis of the common flexor origin), carpal tunnel syndrome.

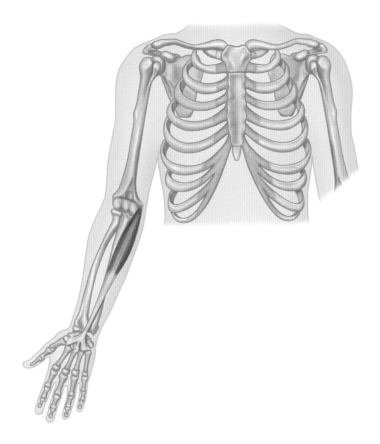

*Anterior view.*

**Latin**, *palmaris*, relating to the palm; *longus*, long.

Part of the superficial layer of the anterior forearm, which also includes pronator teres, flexor carpi radialis, and flexor carpi ulnaris. The palmaris longus muscle is frequently absent.

### Origin
Common flexor origin on anterior aspect of medial epicondyle of humerus.

### Insertion
Superficial (front) surface of flexor retinaculum and apex of palmar aponeurosis.

### Action
Flexes wrist. Tenses palmar fascia.

### Nerve
Median nerve C(6), **7**, **8**, T1.

### Basic functional movement
Examples: grasping a small ball, cupping the palm to drink from the hand.

### Sports that heavily utilize this muscle
Examples: sailing, waterskiing, golf, baseball, cricket, volleyball.

### Movements or injuries that may damage this muscle
Over-extending the wrist as a result of breaking a fall with the hand.

### Common problems when muscle is chronically tight/shortened/overused
Golfer's elbow (overuse tendonitis of the common flexor origin), carpal tunnel syndrome.

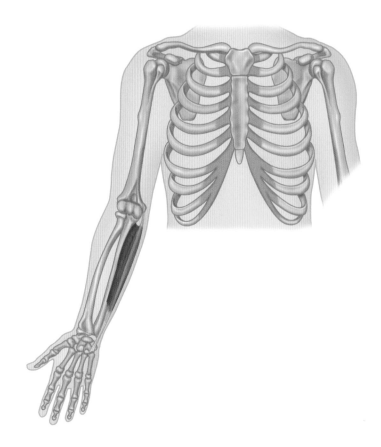

*Anterior view.*

**Latin**, *flectere*, to bend; *carpi*, of the wrist; *ulnaris*, relating to the elbow/arm.

Part of the superficial layer of the anterior forearm, which also includes pronator teres, flexor carpi radialis, and palmaris longus.

**Origin**
Humeral head: common flexor origin on medial epicondyle of humerus.
Ulnar head: medial border of olecranon. Posterior border of upper two-thirds of ulna.

**Insertion**
Pisiform bone. Hook of hamate. Base of fifth metacarpal.

**Action**
Flexes and adducts wrist. May weakly assist in flexion of elbow.

**Nerve**
Ulnar nerve C7, **8**, T1.

**Basic functional movement**
Example: pulling an object toward you.

**Sports that heavily utilize this muscle**
Examples: sailing, waterskiing, golf, baseball, cricket, volleyball.

**Movements or injuries that may damage this muscle**
Over-extending the wrist as a result of breaking a fall with the hand.

**Common problems when muscle is chronically tight/shortened**
Golfer's elbow (overuse tendonitis of the common flexor origin), carpal tunnel syndrome.

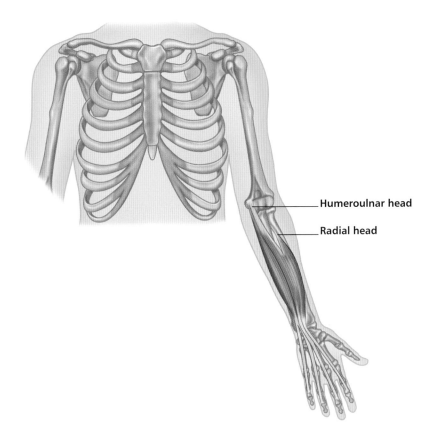

Humeroulnar head

Radial head

*Anterior view.*

**Latin**, *flectere*, to bend; *digitorum*, of the fingers/toes; *superficialis*, on the surface.

This muscle alone constitutes the middle layer of the anterior forearm muscles.

## Origin
Humeroulnar head: long linear origin from common flexor tendon on medial epicondyle of humerus. Medial border of coronoid process of ulna.
Radial head: upper two-thirds of anterior border of radius.

## Insertion
Four tendons each divide into two slips, each of which insert into the sides of the middle phalanges of the four fingers.

## Action
Flexes middle phalanges of each finger. Can help flex wrist.

## Nerve
Median nerve C**7**, **8**, T**1**.

## Basic functional movement
Examples: "hook grip," "power grip" (as in turning a tap), typing, playing the piano and some stringed instruments.

## Sports that heavily utilize this muscle
Examples: archery, maintaining grip in racket and batting sports, judo, rowing, rock-face climbing.

## Movements or injuries that may damage this muscle
Over-extending the wrist as a result of breaking a fall with the hand.

## Common problems when muscle is chronically tight/shortened/overused
Golfer's elbow (overuse tendonitis of the common flexor origin). Carpal tunnel syndrome.

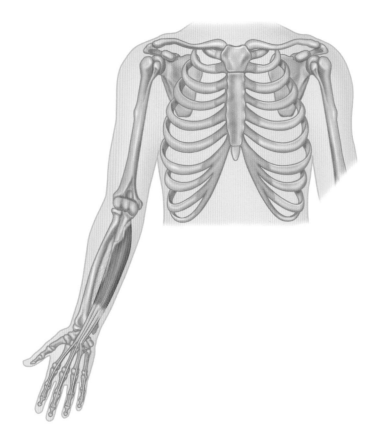

*Anterior view.*

**Latin**, *flectere*, to bend; *digitorum*, of the fingers/toes; *profundus*, deep.

Part of the deep layer (third layer) of the anterior forearm, which also includes flexor pollicis longus and pronator quadratus. In the palm, the tendons of flexor digitorum profundus provide the origin for lumbricales.

## Origin
Upper two-thirds of medial and anterior surfaces of ulna, reaching up onto medial side of olecranon process. Interosseous membrane.

## Insertion
Anterior surface of base of distal phalanges.

## Action
Flexes distal phalanges (the only muscle able to do so). Helps flex all joints across which it passes.

## Nerve
Medial half of muscle, destined for the little and ring fingers: ulnar nerve C7, **8**, **T1**.
Lateral half of muscle, destined for the index and middle fingers: anterior interosseous branch of median nerve C7, **8**, **T1**. Sometimes the ulnar nerve supplies the whole muscle.

## Basic functional movement
Example: "hook grip," as in carrying a briefcase.

## Sports that heavily utilize this muscle
Examples: archery, maintaining grip in racket and batting sports, judo, rowing, rock-face climbing.

## Movements or injuries that may damage this muscle
Over-extending the wrist as a result of breaking a fall with the hand.

## Common problems when muscle is chronically tight/ shortened (spastic)
Golfer's elbow (overuse tendonitis of the common flexor origin). Carpal tunnel syndrome.

# FLEXOR POLLICIS LONGUS

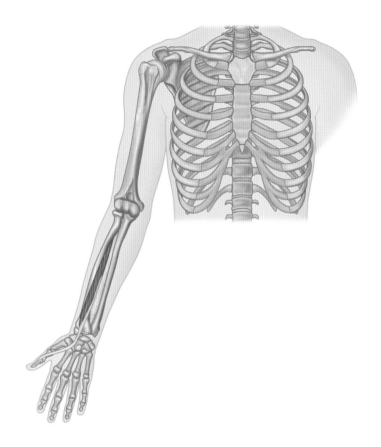

*Anterior view.*

**Latin**, *flectere*, to bend; *pollicis*, of the thumb; *longus*, long.

Part of the deep layer (third layer) of the anterior forearm, which also includes flexor digitorum profundus and pronator quadratus. The tendon of flexor pollicis longus, along with the other long digital flexor tendons, passes through the carpal tunnel.

## Origin
Middle part of anterior surface of shaft of radius. Interosseous membrane. Medial border of coronoid process of ulna and/or medial epicondyle of humerus.

## Insertion
Palmar surface of base of distal phalanx of thumb.

## Action
Flexes interphalangeal joint of thumb (the only muscle able to do so). Assists in flexion of metacarpophalangeal and carpometacarpal joints. Can assist in flexion of wrist.

## Nerve
Anterior interosseous branch of median nerve C(6), 7, **8**, **T1**.

## Basic functional movement
Examples: picking up small objects between the thumb and fingers, maintaining a firm grip on a hammer.

## Sports that heavily utilize this muscle
Examples: archery, maintaining grip in racket and batting sports, judo, rowing, rock-face climbing.

## Movements or injuries that may damage this muscle
Over-extending the wrist as a result of breaking a fall with the hand.

## Common problems when muscle is chronically tight/ shortened/overused
Golfer's elbow (overuse tendonitis of the common flexor origin). Carpal tunnel syndrome.

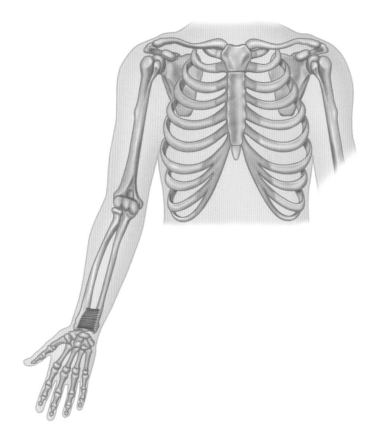

*Anterior view.*

**Latin**, *pronare*, to bend forward; *quadratus*, squared.

Part of the deep layer (third layer) of the anterior forearm, which also includes flexor digitorum profundus and flexor pollicis longus.

**Origin**
Distal quarter of anterior surface of shaft of ulna.

**Insertion**
Lateral side of distal quarter of anterior surface of shaft of radius.

**Action**
Pronates forearm and hand. Helps hold radius and ulna together, reducing stress on inferior radioulnar joint.

**Nerve**
Anterior interosseous branch of median nerve C7, **8**, T1.

**Basic functional movement**
Example: turning the hand downward, as in pouring a substance out of the hand.

**Sports that heavily utilize this muscle**
Examples: archery, maintaining grip in racket and batting sports, judo, rowing, rock-face climbing.

**Movements or injuries that may damage this muscle**
Over-extending the wrist as a result of breaking a fall with the hand.

**Common problems when muscle is chronically tight/ shortened (spastic)**
Golfer's elbow (overuse tendonitis of the common flexor origin). Carpal tunnel syndrome.

# Muscles of the Posterior Forearm

On the back of the forearm there are two muscle groups—superficial and deep.

**Strengthen**

Wrist-weighted rotations     Dumbbell overhead extension     Reverse wrist curls

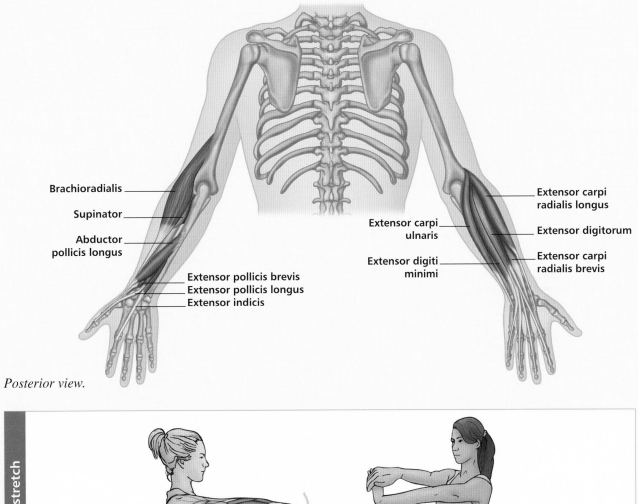

Brachioradialis

Supinator

Abductor
pollicis longus

Extensor pollicis brevis
Extensor pollicis longus
Extensor indicis

Extensor carpi
radialis longus

Extensor carpi
ulnaris

Extensor digiti
minimi

Extensor digitorum

Extensor carpi
radialis brevis

*Posterior view.*

**Self-stretch**

Fingers-down wrist stretch     Rotating wrist stretch

# BRACHIORADIALIS

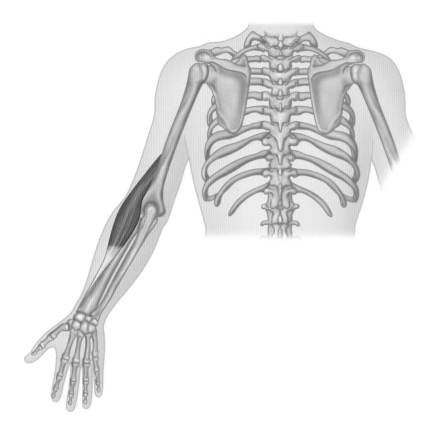

*Posterior view.*

**Latin**, *brachium*, arm; *radius*, staff, spoke of wheel.

Part of the superficial group of the posterior forearm. Brachioradialis forms the lateral border of the cubital fossa. The muscle belly is prominent when working against resistance.

**Origin**
Upper two-thirds of anterior aspect of lateral supracondylar ridge of humerus.

**Insertion**
Lower lateral end of radius, just above styloid process.

**Action**
Flexes elbow joint. Assists in pronation and supination of forearm when these movements are resisted.

**Nerve**
Radial nerve C**5**, **6**.

**Basic functional movement**
Example: turning a corkscrew.

**Sports that heavily utilize this muscle**
Examples: baseball, cricket, golf, racket sports, rowing.

# EXTENSOR CARPI RADIALIS LONGUS

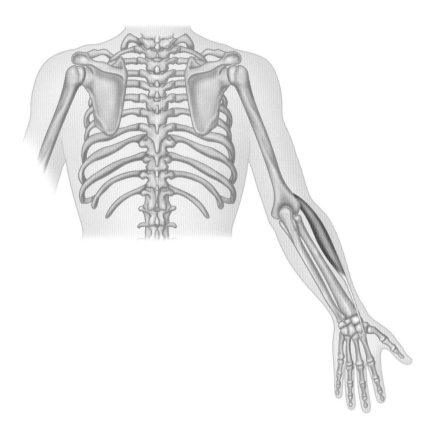

*Posterior view.*

**Latin**, *extendere*, to extend; *carpi*, of the wrist; *radius*, staff, spoke of wheel; *longus*, long.

Part of the superficial group of the posterior forearm. The fibers of this muscle are often blended with those of brachioradialis.

### Origin
Lower (distal) third of lateral supracondylar ridge of humerus.

### Insertion
Dorsal surface of base of second metacarpal bone, on its radial side.

### Action
Extends and abducts wrist. Assists in flexion of elbow.

### Nerve
Radial nerve C5, **6**, **7**, 8.

### Basic functional movement
Examples: kneading dough, typing.

### Sports that heavily utilize this muscle
Examples: backhand in badminton, golf, motorcycle sports (throttle control).

### Movements or injuries that may damage this muscle
Over-flexing the wrist as a result of falling onto the hand.

### Common problems when muscle is chronically tight/ shortened (spastic)
Tennis elbow (overuse tendonitis of the common origin on the lateral epicondyle of the humerus).

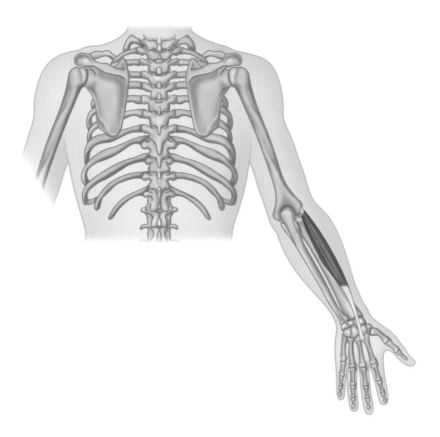

*Posterior view.*

**Latin**, *extendere*, to extend; *carpi*, of the wrist; *radius*, staff, spoke of wheel; *brevis*, short.

Part of the superficial group of the posterior forearm. This muscle is often fused with extensor carpi radialis longus at its origin.

**Origin**
Common extensor tendon from lateral epicondyle of humerus.

**Insertion**
Dorsal surface of third metacarpal.

**Action**
Extends wrist. Assists in abduction of wrist.

**Nerve**
Radial nerve C5, **6**, **7**, 8.

**Basic functional movement**
Examples: kneading dough, typing.

**Sports that heavily utilize this muscle**
Examples: backhand in badminton, golf, motorcycle sports (throttle control).

**Movements or injuries that may damage this muscle**
Over-flexing the wrist as a result of falling onto the hand.

**Common problems when muscle is chronically tight/ shortened (spastic)**
Tennis elbow (overuse tendonitis of the common origin on the lateral epicondyle of the humerus).

# EXTENSOR DIGITORUM

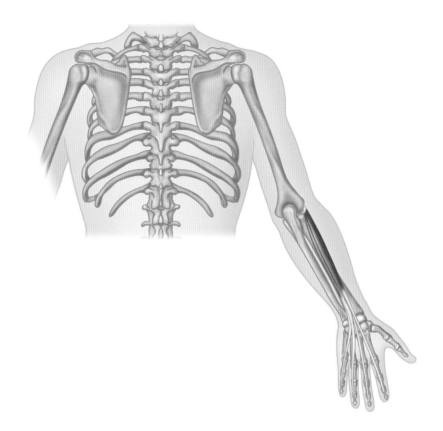

*Posterior view.*

**Latin**, *extendere*, to extend; *digitorum*, of the fingers/toes.

Part of the superficial group of the posterior forearm. Each tendon of extensor digitorum, over each metacarpophalangeal joint, forms a triangular membranous sheet called the *extensor hood* or *extensor expansion*, into which inserts the lumbricales and interossei of the hand. Extensor digiti minimi and extensor indicis also insert into the extensor expansion.

**Origin**
Common extensor tendon from lateral epicondyle of humerus.

**Insertion**
Dorsal surfaces of all the phalanges of the four fingers.

**Action**
Extends fingers (metacarpophalangeal and interphalangeal joints). Assists in abduction (divergence) of fingers away from middle finger.

**Nerve**
Deep radial (posterior interosseous) nerve C**6**, **7**, **8**.

**Basic functional movement**
Example: letting go of objects held in the hand.

**Movements or injuries that may damage this muscle**
Over-flexing the wrist as a result of falling onto the hand.

**Common problems when muscle is chronically tight/ shortened (spastic)**
Tennis elbow (overuse tendonitis of the common origin on the lateral epicondyle of the humerus).

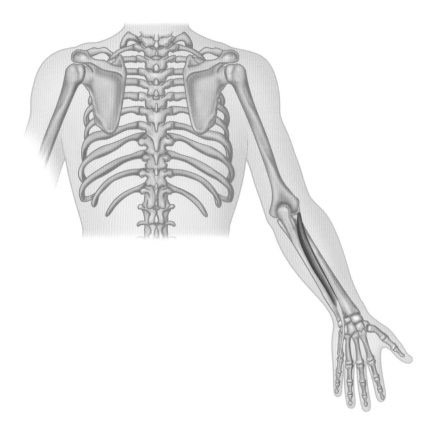

*Posterior view.*

**Latin**, *extendere*, to extend; *digiti*, of the finger/toe; *minimi*, of the smallest.

Part of the superficial group of the posterior forearm, along with brachioradialis, extensor carpi radialis longus, extensor carpi radialis brevis, extensor digitorum, and extensor carpi ulnaris.

**Origin**
Common extensor tendon from lateral epicondyle of humerus.

**Insertion**
Extensor expansion of little finger with extensor digitorum tendon.

**Action**
Extends little finger.

**Nerve**
Deep radial (posterior interosseous) nerve C6, **7**, **8**.

# EXTENSOR CARPI ULNARIS

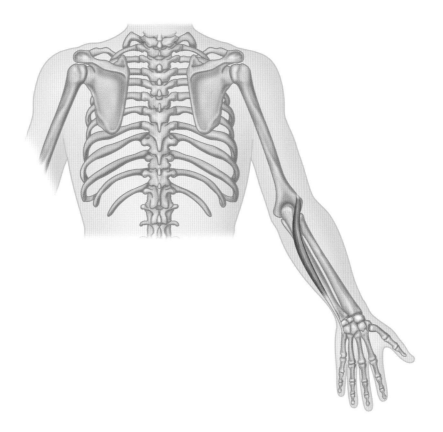

*Posterior view.*

**Latin**, *extendere*, to extend; *carpi*, of the wrist; *ulnaris*, relating to the elbow/arm.

Part of the superficial group of the posterior forearm, along with brachioradialis, extensor carpi radialis longus, extensor carpi radialis brevis, extensor digitorum, and extensor digiti minimi.

### Origin
Common extensor tendon from lateral epicondyle of humerus. Aponeurosis from mid-posterior border of ulna.

### Insertion
Medial side of base of fifth metacarpal.

### Action
Extends and adducts wrist.

### Nerve
Deep radial (posterior interosseous) nerve C6, **7**, **8**.

### Basic functional movement
Example: cleaning windows.

### Sports that heavily utilize this muscle
Examples: backhand in badminton, golf, motorcycle sports (throttle control).

### Movements or injuries that may damage this muscle
Over-flexing the wrist as a result of falling onto the hand.

### Common problems when muscle is chronically tight/shortened (spastic)
Tennis elbow (overuse tendonitis of the common origin on the lateral epicondyle of the humerus).

# SUPINATOR

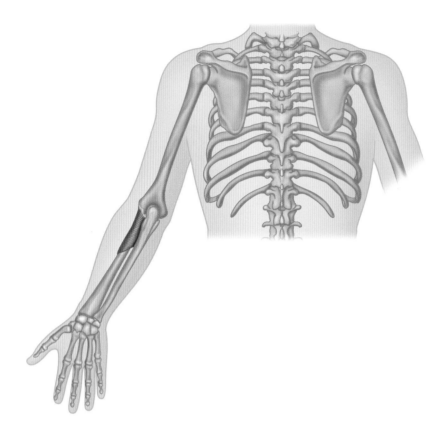

*Posterior view.*

**Latin**, *supinus*, lying on the back.

Part of the deep group of muscles of the posterior forearm. Supinator is almost entirely concealed by the superficial muscles.

**Origin**
Lateral epicondyle of humerus. Radial collateral (lateral) ligament of elbow joint. Annular ligament of superior radioulnar joint. Supinator crest of ulna.

**Insertion**
Dorsal and lateral surfaces of upper third of radius.

**Action**
Supinates forearm (for which it is probably the main prime mover, with biceps brachii being an auxiliary muscle).

**Nerve**
Posterior interosseous nerve, a continuation of the deep branch of radial nerve C5, **6**, (7).

**Basic functional movement**
Example: turning a door handle or screwdriver.

**Sports that heavily utilize this muscle**
Example: backhand in racket sports.

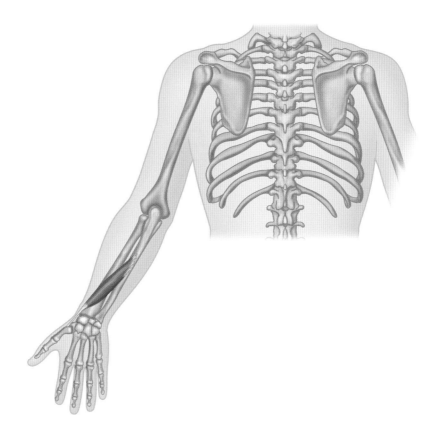

*Posterior view.*

**Latin**, *abducere*, to lead away from; *pollicis*, of the thumb; *longus*, long.

Part of the deep group of muscles of the posterior forearm. However, abductor pollicis longus becomes superficial in the distal part of the forearm.

**Origin**
Posterior surface of shaft of ulna, distal to origin of supinator. Interosseous membrane. Posterior surface of middle third of shaft of radius.

**Insertion**
Radial (lateral) side of base of first metacarpal.

**Action**
Pulls metacarpal bone of thumb into a position midway between extension and abduction (the tendon stands out during this movement). Abducts and assists in flexion of wrist.

**Nerve**
Deep radial (posterior interosseous) nerve C6, **7**, **8**.

**Basic functional movement**
Example: releasing the grip on a flat object.

# EXTENSOR POLLICIS BREVIS

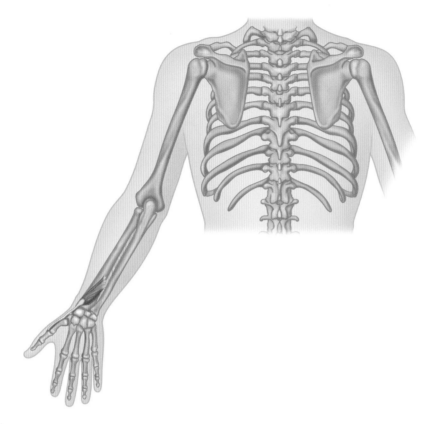

*Posterior view.*

**Latin**, *extendere*, to extend; *pollicis*, of the thumb; *brevis*, short.

Part of the deep group of muscles of the posterior forearm. Extensor pollicis brevis lies distal to abductor pollicis longus, to which it closely adheres.

**Origin**
Posterior surface of radius, distal to origin of abductor pollicis longus. Adjacent part of interosseous membrane.

**Insertion**
Base of dorsal surface of proximal phalanx of thumb.

**Action**
Extends thumb. Abducts wrist.

**Nerve**
Deep radial (posterior interosseous) nerve C6, **7**, **8**.

**Basic functional movement**
Example: releasing the grip on a flat object.

# EXTENSOR POLLICIS LONGUS

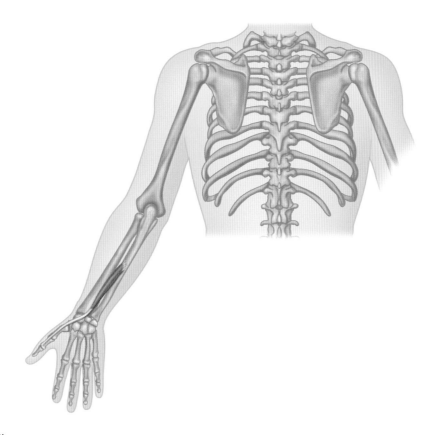

*Posterior view.*

**Latin**, *extendere*, to extend; *pollicis*, of the thumb; *longus*, long.

Part of the deep group of muscles of the posterior forearm. The tendon of extensor pollicis longus forms the posterior boundary of the triangular hollow known as the *anatomical snuffbox*, on the back of the hand, distal to the distal end of the radius.

**Origin**
Middle third of posterior surface of ulna. Interosseous membrane.

**Insertion**
Dorsal surface of base of distal phalanx of thumb.

**Action**
Extends thumb. Assists in extension and abduction of wrist.

**Nerve**
Deep radial (posterior interosseous) nerve C6, **7**, **8**.

**Basic functional movement**
Example: giving the "thumbs up" gesture.

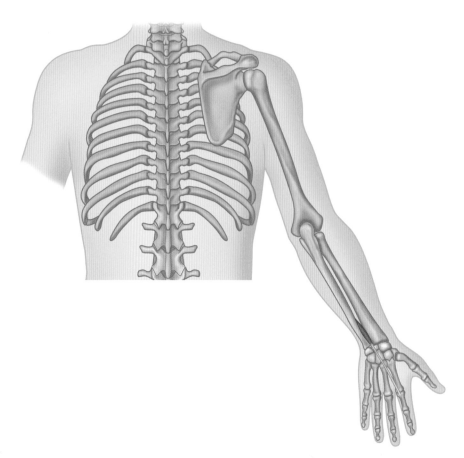

*Posterior view.*

Latin, *extendere*, to extend; *indicis*, of the index finger.

Part of the deep group of muscles of the posterior forearm, which also includes supinator, abductor pollicis longus, extensor pollicis brevis, and extensor pollicis longus.

**Origin**
Posterior surface of ulna. Adjacent part of interosseous membrane.

**Insertion**
Extensor expansion (hood) on dorsum of proximal phalanx of index finger.

**Action**
Extends index finger.

**Nerve**
Deep radial (posterior interosseous) nerve C6, **7**, **8**.

**Basic functional movement**
Example: pointing at something.

# Muscles of the Hand

The muscle groupings in the hand are: 1) the intrinsic muscles, consisting of lumbricales and palmar and dorsal interossei; 2) the muscles of the hypothenar eminence; 3) the muscles of the thenar eminence; and 4) adductor pollicis.

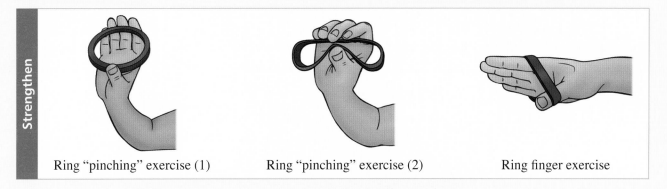

Ring "pinching" exercise (1)     Ring "pinching" exercise (2)     Ring finger exercise

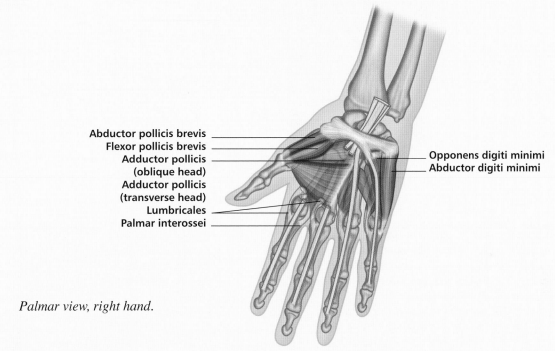

Abductor pollicis brevis
Flexor pollicis brevis
Adductor pollicis
(oblique head)
Adductor pollicis
(transverse head)
Lumbricales
Palmar interossei

Opponens digiti minimi
Abductor digiti minimi

*Palmar view, right hand.*

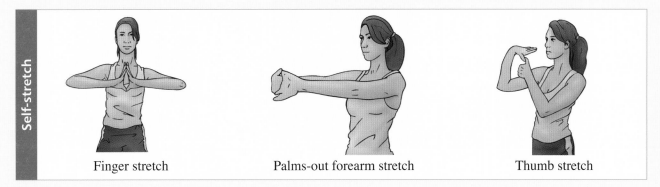

Finger stretch     Palms-out forearm stretch     Thumb stretch

# LUMBRICALES

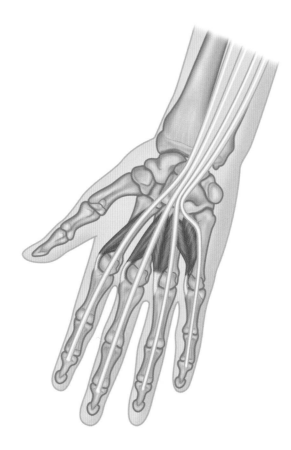

*Palmar view, right hand.*

**Latin**, *lumbricus*, earthworm.

Four small cylindrical muscles, one for each finger, named after the earthworm because of its shape.

**Origin**
Tendons of flexor digitorum profundus in palm.

**Insertion**
Lateral (radial) side of corresponding tendon of extensor digitorum, on dorsum of respective digits.

**Action**
Extend interphalangeal joints and simultaneously flex metacarpophalangeal joints of fingers.

**Nerve**
This varies, but the usual configuration is:
Lateral lumbricales (first and second): median nerve C(6), 7, **8**, T1.
Medial lumbricales (third and fourth): ulnar nerve C(7), **8**, T1. However, the number of lumbricales supplied by the ulnar nerve may be increased to four or decreased to one.

**Basic functional movement**
Example: cupping the hand.

**Sports that heavily utilize these muscles**
Examples: volleyball, handball.

**Common problems when muscles are chronically tight/ shortened (spastic)**
Clawed hand. Inability to maintain flexion of the interphalangeal joints, as in rock-face climbing.

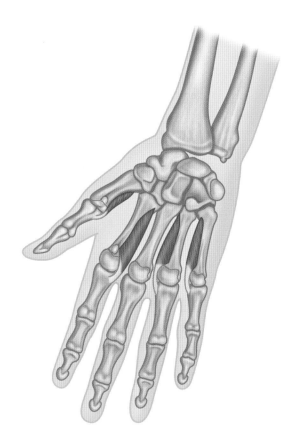

*Palmar view, right hand.*

**Latin**, *palmaris*, relating to the palm; *interosseus*, between bones.

The four palmar interossei are located in the spaces between the metacarpals. Each muscle arises from the metacarpal of the digit upon which it acts.

## Origin
First: medial (ulnar) side of base of first metacarpal.
Second: medial (ulnar) side of shaft of second metacarpal.
Third: lateral (radial) side of shaft of fourth metacarpal.
Fourth: lateral (radial) side of shaft of fifth metacarpal.

## Insertion
Primarily into extensor expansion of respective digit, with possible attachment to base of proximal phalanx as follows:
First: medial (ulnar) side of proximal phalanx of thumb.
Second: medial (ulnar) side of proximal phalanx of index finger.
Third: lateral (radial) side of proximal phalanx of ring finger.
Fourth: lateral (radial) side of proximal phalanx of little finger.

## Action
Adduct (converge) fingers and thumb toward middle (third) finger. Assist in flexion of fingers at metacarpophalangeal joints.

## Nerve
Ulnar nerve C8, T1.

## Basic functional movement
Example: cupping the hand as if to retain water in the palm (i.e. drinking from the hand).

## Sports that heavily utilize these muscles
Example: rock-face climbing.

Note: The palmar interosseous of the thumb is usually absent.

# DORSAL INTEROSSEI

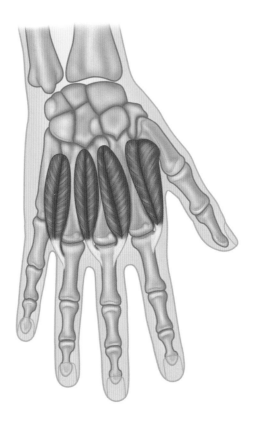

*Dorsal view, right hand.*

**Latin**, *dorsalis*, relating to the back; *interosseus*, between bones.

The four dorsal interossei are about twice the size of the palmar interossei.

**Origin**
By two heads, each from adjacent sides of metacarpals. Therefore, each of the dorsal interossei occupies an interspace between adjacent metacarpals.

**Insertion**
Into extensor expansion and to base of proximal phalanx as follows:
First: lateral (radial) side of index finger, mainly to base of proximal phalanx.
Second: lateral (radial) side of middle finger.
Third: medial (ulnar) side of middle finger, mainly into extensor expansion.
Fourth: medial (ulnar) side of ring finger.

**Action**
Abduct fingers away from middle finger. Assist in flexion of fingers at metacarpophalangeal joints.

**Nerve**
Ulnar nerve C8, T1.

**Basic functional movement**
Example: spreading the fingers as if to indicate numbers from two to four.

**Sports that heavily utilize these muscles**
Example: rock-face climbing.

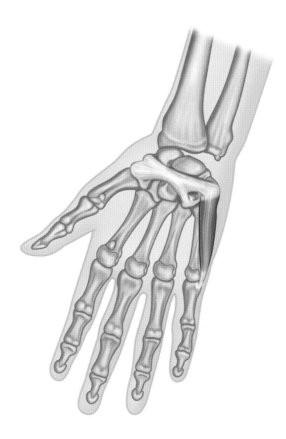

*Palmar view, right hand.*

**Latin**, *abducere*, to lead away from; *digiti*, of the finger/toe; *minimi*, of the smallest.

Most superficial muscle of the hypothenar eminence; other muscles are flexor digiti minimi brevis and opponens digiti minimi.

**Origin**
Pisiform bone. Tendon of flexor carpi ulnaris.

**Insertion**
Ulna (medial) side of base of proximal phalanx of little finger.

**Action**
Abducts little finger. A surprisingly powerful muscle, which particularly comes into play when fingers are spread to grasp a large object.

**Nerve**
Ulnar nerve C(7), **8**, T**1**.

**Basic functional movement**
Example: holding a large ball.

**Sports that heavily utilize this muscle**
Examples: rock-face climbing, basketball, netball.

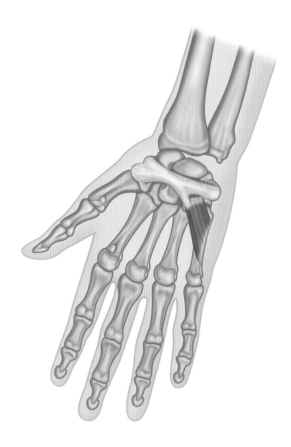

*Palmar view, right hand.*

**Latin**, *opponens*, opposing; *digiti*, of the finger/toe; *minimi*, of the smallest.

Part of the hypothenar eminence, lying deep to abductor digiti minimi.

## Origin
Hook of hamate. Anterior surface of flexor retinaculum.

## Insertion
Entire length of medial (ulnar) border of fifth metacarpal.

## Action
Pulls metacarpal of little finger forward and rotates it laterally, thus deepening hollow of hand and enabling pad of little finger to contact pad of thumb.

## Nerve
Ulnar nerve C(7), **8**, T1.

## Basic functional movement
Example: holding a thread within the fingertips (along with the other fingertips).

## Sports that heavily utilize this muscle
Examples: volleyball, handball, rock-face climbing.

## Common problems when muscle is chronically tight/shortened (spastic)
Over-abducting (opponens digiti minimi) or over-extending (flexor digiti minimi brevis) the little finger as a result of falling onto the ulnar side of the hand.

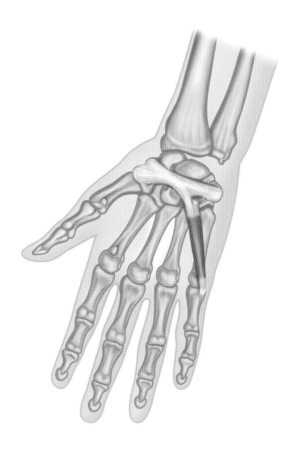

*Palmar view, right hand.*

**Latin**, *flectere*, to flex; *digiti*, of the finger/toe; *minimi*, of the smallest; *brevis*, short.

Part of the hypothenar eminence. May be absent or fused with a neighboring muscle.

### Origin
Hook of hamate. Anterior surface of flexor retinaculum.

### Insertion
Ulna (medial) side of base of proximal phalanx of little finger.

### Action
Flexes little finger at metacarpophalangeal joint.

### Nerve
Ulnar nerve C(7), **8**, T**1**.

### Basic functional movement
Example: holding a thread within the fingertips (along with the other fingertips).

### Sports that heavily utilize this muscle
Examples: volleyball, handball, rock-face climbing.

### Common problems when muscle is chronically tight/shortened (spastic)
Over-abducting (opponens digiti minimi) or over-extending (flexor digiti minimi brevis) the little finger as a result of falling onto the ulnar side of the hand.

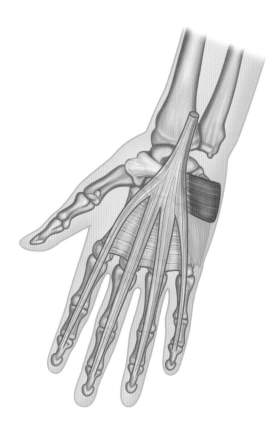

*Palmar view, right hand.*

**Latin**, *palmaris*, relating to the palm; *brevis*, short.

A small subcutaneous muscle lying over the hypothenar eminence.

**Origin**
Palmar aponeurosis. Flexor retinaculum.

**Insertion**
Skin on ulnar border of hand.

**Action**
Wrinkles skin on ulnar border of hand.

**Nerve**
Ulnar nerve C(7), **8**, T1.

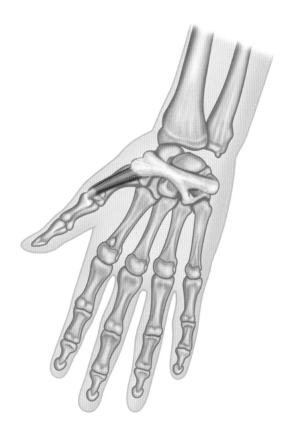

*Palmar view, right hand.*

**Latin**, *abducere*, to lead away from; *pollicis*, of the thumb; *brevis*, short.

Most superficial of the muscles of the thenar eminence; other muscles are flexor pollicis brevis and opponens pollicis.

**Origin**
Flexor retinaculum. Tubercle of trapezium. Tubercle of scaphoid.

**Insertion**
Radial side of base of proximal phalanx of thumb.

**Action**
Abducts thumb and moves it anteriorly (as in typing or playing the piano). Assists in opposition of thumb.

**Nerve**
Median nerve (C6, 7, 8, T1).

**Basic functional movement**
Example: typing.

**Sports that heavily utilize this muscle**
Example: rock-face climbing.

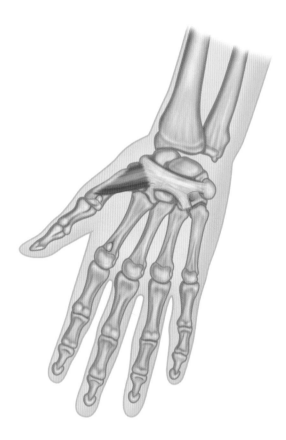

*Palmar view, right hand.*

**Latin**, *opponens*, opposing; *pollicis*, of the thumb.

Part of the thenar eminence. Usually partly fused with flexor pollicis brevis and deep to abductor pollicis brevis.

**Origin**
Flexor retinaculum. Tubercle of trapezium.

**Insertion**
Entire length of radial border of first metacarpal.

**Action**
Opposes (i.e. abducts, then slightly medially rotates, followed by flexion and adduction) thumb so that pad of thumb can be drawn into contact with pads of fingers.

**Nerve**
Median nerve (C6, 7, 8, T1)

**Basic functional movement**
Example: picking up a small object between the thumb and fingers.

**Sports that heavily utilize this muscle**
Examples: rock-face climbing, motorcycle sports (clutch and throttle movement).

**Movements or injuries that may damage this muscle**
Over-extending the thumb as a result of falling onto the hand (rare).

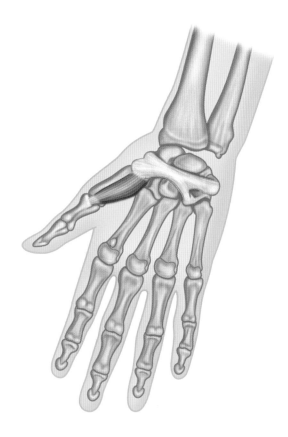

*Palmar view, right hand.*

**Latin**, *flectere*, to flex; *pollicis*, of the thumb; *brevis*, short.

Part of the thenar eminence, along with opponens pollicis (to which it is usually partly fused) and abductor pollicis brevis.

**Origin**
Superficial head: flexor retinaculum; trapezium.
Deep head: trapezoid; capitate.

**Insertion**
Radial side of base of proximal phalanx of thumb.

**Action**
Flexes metacarpophalangeal and carpometacarpal joints of thumb. Assists in opposition of thumb toward little finger.

**Nerve**
Superficial head: median nerve (C6, 7, 8, T1).
Deep head: ulnar nerve (C**8**, T1).

**Basic functional movement**
Example: holding a thread between the thumb and fingertips.

**Sports that heavily utilize this muscle**
Examples: rock-face climbing, motorcycle sports (clutch and throttle movement).

**Movements or injuries that may damage this muscle**
Over-extending the thumb as a result of falling onto the hand (rare).

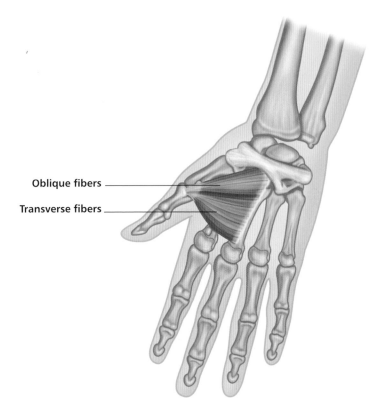

Oblique fibers

Transverse fibers

*Palmar view, right hand.*

**Latin**, *adducere*, to lead to; *pollicis*, of the thumb.

## Origin
Oblique fibers: anterior surfaces of second and third metacarpals, capitate, and trapezoid.
Transverse fibers: palmar surface of third metacarpal bone.

## Insertion
Ulna (medial) side of base of proximal phalanx of thumb.

## Action
Adducts thumb.

## Nerve
Deep ulnar nerve **C8**, **T1**.

## Basic functional movement
Example: gripping a jam jar lid to screw it on.

## Sports that heavily utilize this muscle
Example: rock-face climbing.

## Movements or injuries that may damage this muscle
Over-abducting the thumb as a result of falling onto the hand.

# 8  Muscles of the Hip and Thigh

The hips and buttocks are comprised of a number of both large (e.g. gluteus maximus) and small (e.g. piriformis) muscles; these muscles are primarily responsible for hip stabilization and lower limb movement. The muscles around the hip and buttocks, along with the structure of the hip joint, allow a large range of motion of the lower limb, including flexion, extension, adduction, abduction, and rotation.

The bulk of the buttock is formed mainly by the **gluteus maximus**, which is the largest and most superficial muscle of the group, lying posterior to smaller muscles, such as gluteus medius and gluteus minimus. Gluteus maximus contributes to powerful hip extension for explosive activities such as sprinting.

The abductors (tensor fasciae latae, gluteus medius, and gluteus minimus) are located on the lateral side (outside) of the thigh and posterior pelvis. They originate at the top outer edge of the pelvis and extend down the outside of the thigh, attaching to the lateral side of the tibia. The primary action of the abductors is to abduct (draw away from the midline) and medially (internally) rotate the hip joint.

**Tensor fasciae latae** lies anterior to gluteus maximus. It is a superficial muscle of the upper thigh that keeps the pelvis level and stabilizes the knee when standing on one leg; it also assists in flexion of the hip joint. Many painful conditions of the knee can arise because of a short spastic tensor fasciae latae.

**Gluteus medius** is mostly deep to, and therefore obscured by, gluteus maximus, but appears on the pelvic surface between gluteus maximus and tensor fasciae latae. During walking, gluteus medius, along with gluteus minimus, prevents the pelvis from dropping toward the non-weight-bearing leg. When gluteus medius is tight, pelvic imbalances may result, leading to pain in the hips, low back, and knees. **Gluteus minimus** lies deep to gluteus medius, whose fibers obscure it; as its name implies, it is the smallest of the gluteal muscles. As with gluteus medius, when

gluteus minimus is tight, pelvic imbalances may occur.

Located under the gluteal muscles, the six deep rotators (piriformis, gemellus superior and inferior, obturator internus and externus, and quadratus femoris) are the smallest muscles of the hip and are primarily responsible for lateral (outward) rotation.

**Piriformis** is a small, tubular muscle that originates on the anterior surface of the sacrum, inserts at the superior border of the greater trochanter of the femur (posterior trochanteric line), and leaves the pelvis by passing through the greater sciatic foramen. The muscle assists in laterally rotating the femur in the hip joint and in abducting the thigh when the hip is flexed; it also helps hold the head of the femur in the acetabulum.

**Gemellus superior** and **gemellus inferior** (the **gemelli**) are small, thin muscles that cross the hip joint from the area of the ischium to the greater trochanter of the femur. Their path is almost horizontal across the joint.

Lying between the two gemelli, the **obturator internus** has a broad origin on a part of the pelvis called the *obturator foramen*, along with portions of the lower iliac bone. Besides being an outward rotator, it is a strong stabilizer of the hip.

The **obturator externus** is an ideal rotator of the hip because of its position. It passes from the lower end of the obturator foramen, then behind the neck of the femur, to attach to the greater trochanter of the femur on the medial side. Its line of pull allows the head of the femur to roll laterally inside the socket of the pelvis, creating outward, or external, rotation.

The most inferior (lowest) deep rotator is the **quadratus femoris**; this is a short muscle running almost horizontally from the upper portion of the ischial tuberosity to the femur (at the level of the gluteal crease).

The hamstrings are a large group of three separate muscles located in the posterior (rear) thigh. Originating from the bottom of the ischial tuberosity and extending to below the knee, they work together to extend the hip and flex the knee; these muscles correspond to the flexors of the elbow in the upper limb. During running, the hamstrings decelerate the leg at the end of its forward swing and prevent the trunk from flexing at the hip joint.

The three muscles of the hamstrings are, from medial to lateral, semimembranosus**,** semitendinosus**,** and biceps femoris. **Biceps femoris** is the largest hamstring muscle, and has two heads, the long and the short; the long head crosses the hip joint to work it. **Semitendinosus** and **semimembranosus** are completely synergistic, meaning that they both do the exact same actions; however, they can also oppose each other, causing internal or external rotation of the tibia–fibula complex.

The adductors are a large group of muscles located on the medial (inner) side of the thigh. They originate at the bottom of the hipbone and extend down the inside of the thigh, attaching to the medial/posterior side of the femur.

**Pectineus** is the most superior adductor; its primary action is adduction, or bringing the thigh toward the midline of the body. **Gracilis** attaches from the pubis symphysis to the tibia, below the knee. The muscle shapes the superficial inner thigh, but is relatively weak, working the knee joint as well as the hip.

The three muscles specifically named *adductors* are **adductor magnus, adductor brevis**, and **adductor longus**. North American anatomists also include an adductor minimis. The muscles travel down the inside of the thigh, starting at the anterior pubis area of the pelvis and attaching to the medial length of the femur. The magnus is the largest of the three, and spreads out to cover the fullest area of the inner thigh.

The primary action of the adductors is to adduct (draw toward the midline) the femur in the hip joint, but most of them also rotate the femur: pectineus and gracilis inwardly rotate, and adductor brevis and magnus outwardly rotate. All adductors function as stabilizers of the lower limb when weight is placed upon it, as well as stabilizers of the pelvis.

The quadriceps (or quadriceps femoris) is a large group of muscles, the heftiest of the lower limb, located in the anterior (front) of the thigh. These muscles originate from above the hip joint and extend to below the knee. The primary action of the quadriceps is to extend the knee joint; however, in conjunction with a number of other muscles in the front of the hip, the quadriceps is also associated with hip flexion.

**Rectus femoris** is part of the quadriceps femoris, which also includes the vasti group (vastus lateralis, vastus medialis, and vastus intermedius). This spindle-shaped bipennate muscle has two heads of origin: the reflected head is in the line of pull of the muscle in four-footed animals, whereas the straight head seems to have developed in humans as a result of the upright posture.

The quadriceps straightens the knee when rising from sitting, during walking, and while climbing. The **vasti** muscles cross only the knee, and thus are limited to knee extension or resistance to knee flexion; they pay out to control the movement of sitting down. **Vastus medialis** is larger and bulkier (also referred to as the *teardrop muscle*) than **vastus lateralis**. **Vastus intermedius** is the deepest part of the quadriceps femoris, and has a membranous tendon on its anterior surface to allow a gliding movement between itself and the rectus femoris, which overlies it. The quadriceps tendon attaches to, and covers, the patella; it becomes the patellar tendon below the knee, attaching to the tibia.

Included here is **sartorius**, not part of the quadriceps femoris group, but the most superficial muscle of the anterior/medial thigh; it is also the longest strap muscle in the body. The medial border of the upper third of this muscle forms the lateral boundary of the femoral triangle (adductor longus forms the medial boundary and the inguinal ligament forms the superior boundary). The action of sartorius is to put the lower limbs in the cross-legged seated position of the tailor (hence its name from the Latin).

# Muscles of the Buttock

The bulk of the buttock is formed mainly by gluteus maximus, which is the largest and most superficial muscle of the group, lying posterior to smaller muscles, such as gluteus medius and gluteus minimus. Tensor fasciae latae is included as the most anterior muscle of the group. Other muscles—such as gemelli, quadratus femoris, obturator internus, and piriformis—are sometimes grouped with the muscles of the buttock, but here they have been dealt with under "Muscles of the Hip" (p. 224).

**Strengthen**

Good mornings          Lying lateral leg raise          Single-leg glute raise hold

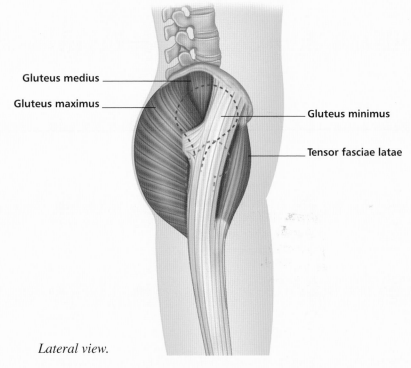

*Lateral view.*

Gluteus medius

Gluteus maximus

Gluteus minimus

Tensor fasciae latae

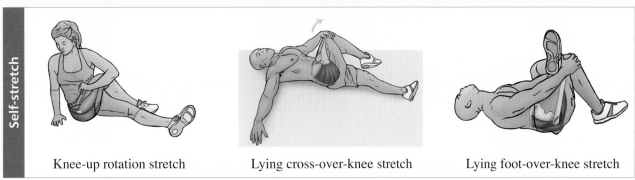

**Self-stretch**

Knee-up rotation stretch          Lying cross-over-knee stretch          Lying foot-over-knee stretch

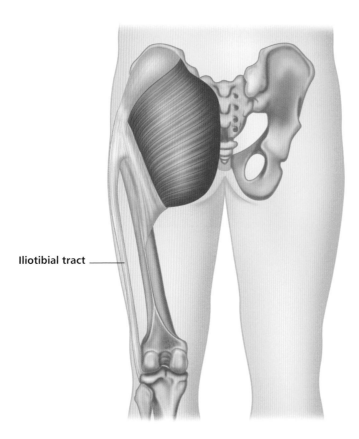

*Posterior view.*

Iliotibial tract

**Greek**, *gloutos*, buttock. **Latin**, *maximus*, biggest.

Gluteus maximus is the most coarsely fibered and heaviest muscle in the body.

### Origin
Outer surface of ilium, behind posterior gluteal line and portion of bone superior and posterior to it. Adjacent posterior surface of sacrum and coccyx. Sacrotuberous ligament. Aponeurosis of erector spinae.

### Insertion
Deep fibers of distal portion: gluteal tuberosity of femur. Remaining fibers: iliotibial tract of fascia lata.

### Action
Upper fibers: laterally rotate hip joint; may assist in abduction of hip joint.
Lower fibers: extend and laterally rotate hip joint (forceful extension, as in running or rising from sitting); extend trunk; assist in adduction of hip joint.
Through its insertion into the iliotibial tract, gluteus maximus helps to stabilize knee in extension.

### Nerve
Inferior gluteal nerve L5, S1, 2.

### Basic functional movement
Examples: walking upstairs, rising from sitting.

### Sports that heavily utilize this muscle
Examples: running, surfing, windsurfing, jumping, weightlifting ("clean" phase, i.e. lifting weights up from floor).

# TENSOR FASCIAE LATAE

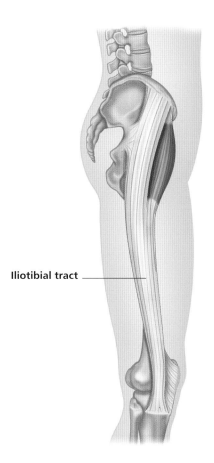

Iliotibial tract

*Lateral view, right leg.*

**Latin**, *tendere*, to stretch, pull; *fasciae*, of the band; *latae*, of the broad.

This muscle lies anterior to gluteus maximus, on the lateral side of the hip.

**Origin**
Anterior part of outer lip of iliac crest, and outer surface of anterior superior iliac spine.

**Insertion**
Joins iliotibial tract just below level of greater trochanter.

**Action**
Flexes, abducts, and medially rotates hip joint. Tenses fascia lata, thus stabilizing knee. Redirects the rotational forces produced by gluteus maximus.

**Nerve**
Superior gluteal nerve L4, 5, S1.

**Basic functional movement**
Example: walking.

**Sports that heavily utilize this muscle**
Examples: horse riding, hurdling, waterskiing.

**Common problems when muscle is chronically tight/ shortened (spastic)**
Pelvic imbalances, leading to pain in the hips, low back, and lateral area of the knees.

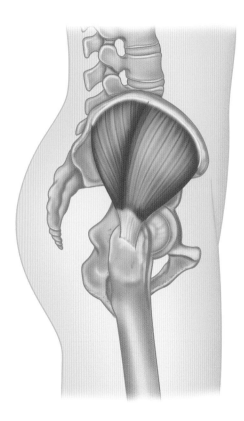

*Lateral view, right leg.*

**Greek**, *gloutos*, buttock. **Latin**, *medius*, middle.

This muscle is mostly deep to, and therefore obscured by, gluteus maximus, but appears on the surface between gluteus maximus and tensor fasciae latae. During walking, gluteus medius, along with gluteus minimus, prevents the pelvis from dropping toward the non-weight-bearing leg.

**Origin**
Outer surface of ilium inferior to iliac crest, between posterior and anterior gluteal lines.

**Insertion**
Oblique ridge on lateral surface of greater trochanter of femur.

**Action**
Abducts hip joint. Anterior fibers medially rotate and may assist in flexion of hip joint. Posterior fibers slightly laterally rotate hip joint.

**Nerve**
Superior gluteal nerve L4, 5, S1.

**Basic functional movement**
Example: stepping sideways over an object, such as a low fence.

**Sports that heavily utilize this muscle**
Examples: all sports requiring side-stepping, especially cross-country skiing and ice skating.

**Common problems when muscle is chronically tight/shortened (spastic)**
Pelvic imbalances, leading to pain in the hips, low back, and knees.

# GLUTEUS MINIMUS

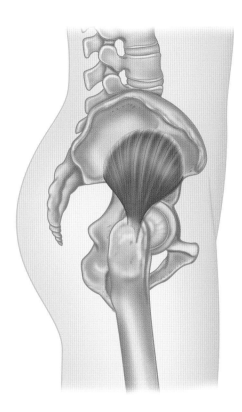

*Lateral view, right leg.*

**Greek**, *gloutos*, buttock. **Latin**, *minimus*, smallest.

This muscle is situated anteroinferior and deep to gluteus medius, whose fibers obscure it.

## Origin
Outer surface of ilium, between anterior and inferior gluteal lines.

## Insertion
Anterior border of greater trochanter.

## Action
Abducts, medially rotates, and may assist in flexion of hip joint.

## Nerve
Superior gluteal nerve L**4**, **5**, S**1**.

## Basic functional movement
Example: stepping sideways over an object, such as a low fence.

## Sports that heavily utilize this muscle
Examples: all sports requiring side-stepping, especially cross-country skiing and ice skating.

## Common problems when muscle is chronically tight/shortened (spastic)
Pelvic imbalances, leading to pain in the hips, low back, and knees.

# Muscles of the Hip

The muscles of the hip are relatively small muscles originating from the sacrum and/or inner surface of the pelvis, to insert on or near the greater trochanter of the femur. They all contribute to lateral rotation of the hip joint. In a role similar to that of the rotator cuff muscles of the shoulder joint, the hip muscles (especially piriformis and obturator internus) also help hold the head of the femur in the acetabulum.

**Strengthen**

Isometric glute squeeze

Low-pulley hip abduction

Resistance-band abduction side steps

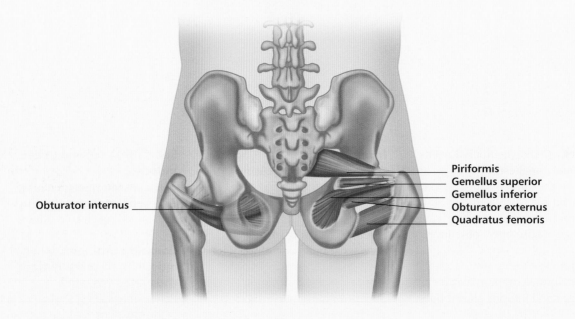

Obturator internus

Piriformis
Gemellus superior
Gemellus inferior
Obturator externus
Quadratus femoris

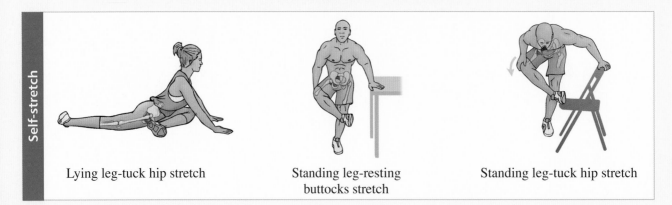

**Self-stretch**

Lying leg-tuck hip stretch

Standing leg-resting buttocks stretch

Standing leg-tuck hip stretch

# PIRIFORMIS

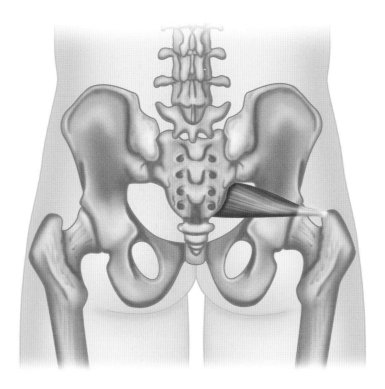

*Posterior view.*

**Latin**, *pirum*, pear; *forma*, shape.

Piriformis leaves the pelvis by passing through the greater sciatic foramen.

**Origin**
Internal surface of sacrum. Sacrotuberous ligament.

**Insertion**
Superior border of greater trochanter of femur.

**Action**
Laterally rotates hip joint. Abducts thigh when hip is flexed. Helps hold head of femur in acetabulum.

**Nerve**
Ventral rami of lumbar nerve L(5) and sacral nerves S**1**, **2**.

**Basic functional movement**
Example: bringing the first leg out of a car.

**Sports that heavily utilize this muscle**
Examples: swimming (breaststroke legs), soccer.

**Common problems when muscle is chronically tight/ shortened (spastic)**
A hypertonic piriformis may squeeze the sciatic nerve, causing piriformis syndrome, i.e. sciatic pain which begins in the buttocks.

# OBTURATOR INTERNUS

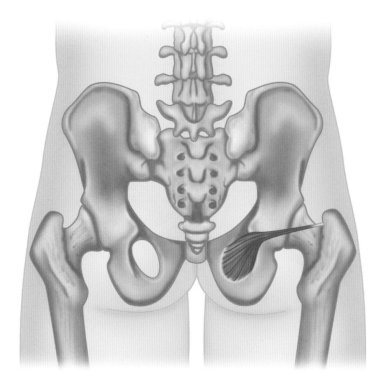

*Posterior view.*

**Latin**, *obturare*, to obstruct; *internus*, internal.

Obturator internus is very closely associated with the two gemelli, in terms of both action and position. It leaves the pelvis by passing through the lesser sciatic foramen.

### Origin
Inner surface of obturator membrane and margin of obturator foramen. Inner surface of ischium, pubis, and ilium.

### Insertion
Medial surface of greater trochanter of femur, above trochanteric fossa.

### Action
Laterally rotates hip joint. Abducts thigh when hip is flexed. Helps hold head of femur in acetabulum.

### Nerve
Nerve to obturator internus, a branch of the ventral rami of lumbar nerve **L5** and sacral nerves **S1, 2**.

### Basic functional movement
Example: bringing the first leg out of a car.

### Sports that heavily utilize this muscle
Examples: swimming (breaststroke legs), soccer.

### Common problems when muscle is chronically tight/ shortened (spastic)
Person stands with the feet turned out.

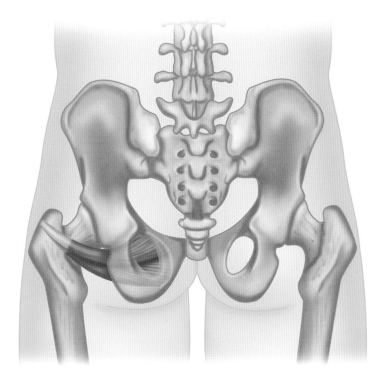

*Posterior view.*

**Latin**, *obturare*, to obstruct; *externus*, external.

This muscle is often grouped with the hip adductors, but is placed in this section because of its similarity and proximity to the other short lateral rotators of the hip.

**Origin**
Rami of pubis and ischium. External surface of obturator membrane.

**Insertion**
Trochanteric fossa of femur.

**Action**
Laterally rotates hip joint. May assist in adduction of hip joint.

**Nerve**
Posterior division of obturator nerve L**3**, **4**.

**Basic functional movement**
Example: clicking the heels together "military style."

**Sports that heavily utilize this muscle**
Examples: swimming (breaststroke legs), soccer.

**Common problems when muscle is chronically tight/ shortened (spastic)**
Person stands with the feet turned out.

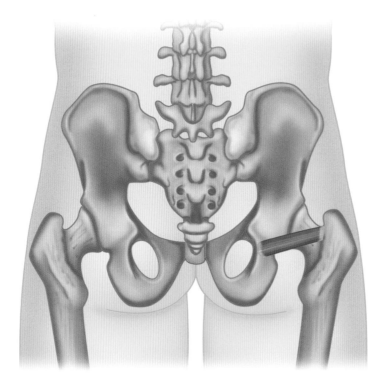

*Posterior view.*

**Latin**, *gemellus*, twin/double; *inferior*, lower.

**Origin**
Upper margin of ischial tuberosity.

**Insertion**
With tendon of obturator internus into medial surface of greater trochanter of femur.

**Action**
Assists obturator internus in lateral rotation of hip joint and abduction of thigh when hip is flexed.

**Nerve**
Branch of nerve to quadratus femoris, a branch of lumbosacral plexus, L**4**, **5**, S**1**, (2).

**Basic functional movement**
Example: bringing the first leg out of a car.

**Sports that heavily utilize this muscle**
Examples: swimming (breaststroke legs), soccer.

**Common problems when muscle is chronically tight/ shortened (spastic)**
Person stands with the feet turned out.

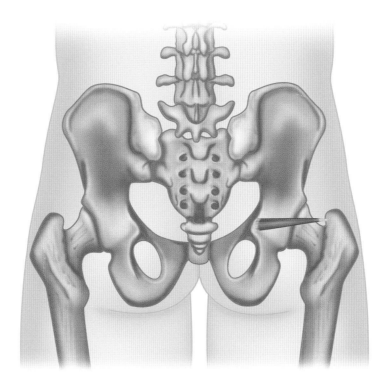

*Posterior view.*

**Latin**, *gemellus*, twin/double; *superior*, upper.

Both gemelli are accessories to obturator internus, providing additional origins from the margins of the lesser sciatic notch.

**Origin**
External surface of ischial spine.

**Insertion**
With tendon of obturator internus into medial surface of greater trochanter of femur.

**Action**
Assists obturator internus in lateral rotation of hip joint and abduction of thigh when hip is flexed.

**Nerve**
Nerve to obturator internus, a branch of the ventral rami of lumbar nerve **L5** and sacral nerves **S1, 2**.

**Basic functional movement**
Example: bringing the first leg out of a car.

**Sports that heavily utilize this muscle**
Examples: swimming (breaststroke legs), soccer.

**Common problems when muscle is chronically tight/ shortened (spastic)**
Person stands with the feet turned out.

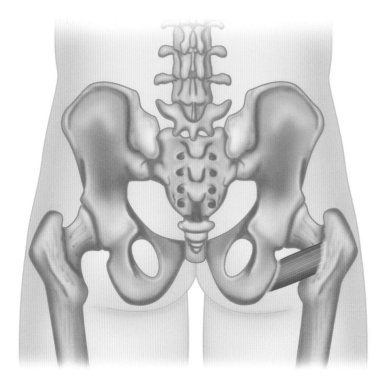

*Posterior view.*

**Latin**, *quadratus*, squared; *femoris*, of the thigh.

This muscle is often fused with either (or both) gemellus inferior, which lies above, and the upper fibers of adductor magnus, which lies below.

**Origin**
Lateral border of ischial tuberosity.

**Insertion**
Quadrate line that extends distally below intertrochanteric crest.

**Action**
Laterally rotates hip joint.

**Nerve**
Nerve to quadratus femoris, a branch of lumbosacral plexus L**4**, **5**, S**1**, (2). This nerve also supplies gemellus inferior.

**Basic functional movement**
Example: bringing the first leg out of a car.

**Sports that heavily utilize this muscle**
Examples: swimming (breaststroke legs), soccer.

**Common problems when muscle is chronically tight/ shortened (spastic)**
Person stands with the feet turned out.

# Muscles of the Thigh

The muscles of the thigh fall broadly into three groups: posterior, medial, and anterior. The posterior thigh consists of the hamstring group, which corresponds to the elbow flexors in the upper limb. The medial thigh consists of the adductor group, which corresponds to coracobrachialis in the upper limb. Obturator externus can also be placed in this group, but has been included under "Muscles of the Hip" (p. 224). The anterior group consists of sartorius along with the four muscles of the quadriceps femoris; this group corresponds to triceps brachii in the upper limb.

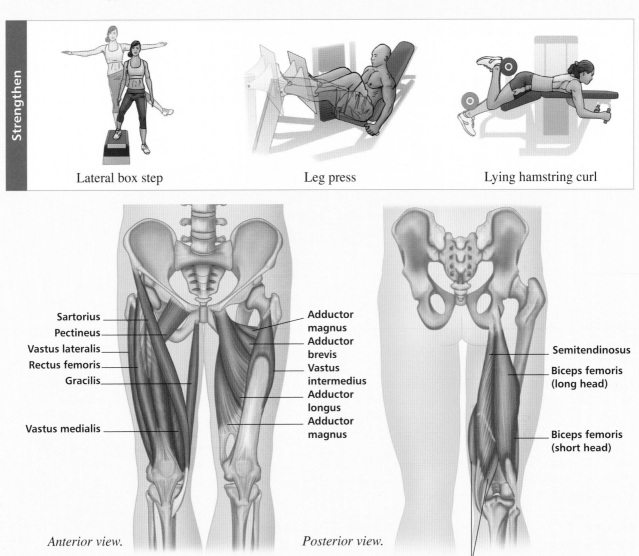

**Strengthen**

Lateral box step

Leg press

Lying hamstring curl

Sartorius

Pectineus

Vastus lateralis

Rectus femoris

Gracilis

Vastus medialis

Adductor magnus

Adductor brevis

Vastus intermedius

Adductor longus

Adductor magnus

*Anterior view.*

Semitendinosus

Biceps femoris (long head)

Biceps femoris (short head)

*Posterior view.*

Semimembranosus

**Self-stretch**

Lying hamstring stretch

Leg-out adductor stretch

Kneeling quad stretch

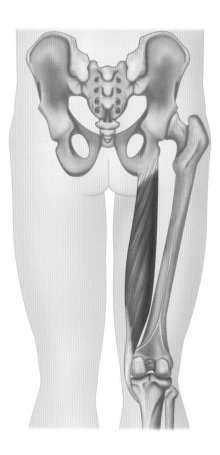

*Posterior view.*

**Latin**, *semi*, half; *tendinosus*, tendinous.

Semitendinosus is the central part of the hamstring group.

**Origin**
Ischial tuberosity.

**Insertion**
Upper medial surface of shaft of tibia.

**Action**
Flexes and slightly medially rotates knee joint after flexion. Extends hip joint.

**Nerve**
Two branches from tibial part of sciatic nerve L**4**, **5**, S**1**, **2**.

**Basic functional movement**
During running, the hamstrings slow down the leg at the end of its forward swing and prevent the trunk from flexing at the hip joint.

**Sports that heavily utilize this muscle**
Examples: sprinting, hurdling, soccer (especially back kicking), jumping and weightlifting (upper portion of hamstrings only).

**Movements or injuries that may damage this muscle**
Sudden lengthening of the muscle without sufficient warm-up (e.g. forward kicking, splits).

**Common problems when muscle is chronically tight/shortened (spastic)**
Low back pain. Knee pain. Leg length discrepancies. Restriction of stride length in walking or running.

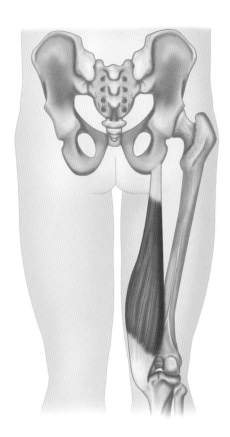

*Posterior view.*

**Latin**, *semi*, half; *membranosus*, membranous.

Medial part of hamstring group. Most of its belly is deep to semitendinosus and the long head of biceps femoris.

### Origin
Ischial tuberosity.

### Insertion
Posteromedial surface of medial condyle of tibia.

### Action
Flexes and slightly medially rotates knee joint after flexion. Extends hip joint.

### Nerve
Two branches from tibial part of sciatic nerve L4, **5**, S**1**, 2.

### Basic functional movement
During running, the hamstrings slow down the leg at the end of its forward swing and prevent the trunk from flexing at the hip joint.

### Sports that heavily utilize this muscle
Examples: sprinting, hurdling, soccer (especially back kicking), jumping and weightlifting (upper portion of hamstrings only).

### Movements or injuries that may damage this muscle
Sudden lengthening of the muscle without sufficient warm-up (e.g. forward kicking, splits).

### Common problems when muscle is chronically tight/ shortened (spastic)
Low back pain. Knee pain. Leg length discrepancies. Restriction of stride length in walking or running.

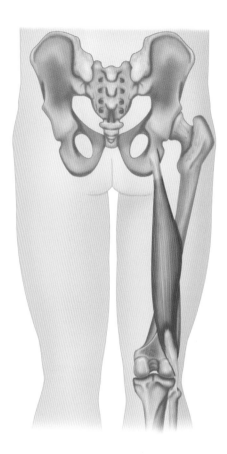

*Posterior view.*

**Latin**, *biceps*, two-headed; *femoris*, of the thigh.

Lateral part of hamstring group.

## Origin
Long head: ischial tuberosity; sacrotuberous ligament.
Short head: linea aspera; upper two-thirds of supracondylar line; lateral intermuscular septum.

## Insertion
Lateral side of head of fibula.
Lateral condyle of tibia.

## Action
Both heads flex knee joint (and laterally rotate the flexed knee joint). Long head also extends hip joint.

## Nerve
Long head: tibial division of sciatic nerve L5, **S1**, **2**, 3.
Short head: peroneal division of sciatic nerve L**5**, **S1**, **2**.

## Basic functional movement
During running, the hamstrings slow down the leg at the end of its forward swing and prevent the trunk from flexing at the hip joint.

## Sports that heavily utilize this muscle
Examples: sprinting, hurdling, soccer (especially back kicking), jumping and weightlifting (upper portion of hamstrings only).

## Movements or injuries that may damage this muscle
Sudden lengthening of the muscle without sufficient warm-up (e.g. forward kicking, splits).

## Common problems when muscle is chronically tight/shortened (spastic)
Low back pain. Knee pain. Leg length discrepancies. Restriction of stride length in walking or running.

# ADDUCTOR MAGNUS

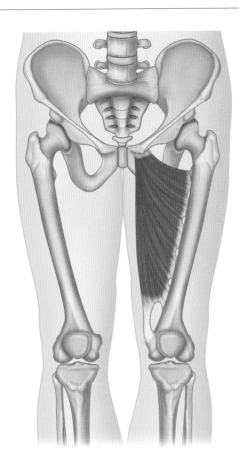

*Anterior view.*

**Latin**, *adducere*, to lead to; *magnus*, large.

Adductor magnus is the largest of the adductor muscle group. Its upper fibers are often fused with those of quadratus femoris. The vertical fibers of the ischial part belong morphologically to the hamstring group, and are therefore supplied by the tibial nerve.

### Origin
Inferior ramus of pubis. Ramus of ischium (anterior fibers). Ischial tuberosity (posterior fibers).

### Insertion
Entire length of femur, along linea aspera and medial supracondylar line to adductor tubercle on medial epicondyle of femur.

### Action
Upper fibers adduct and laterally rotate hip joint. Vertical fibers from ischium may assist in weak extension of hip joint.

### Nerve
Posterior division of obturator nerve L2, **3**, **4**. Tibial division of sciatic nerve L**4**, 5, S1.

### Basic functional movement
Example: bringing the second leg in or out of a car.

### Sports that heavily utilize this muscle
Examples: horse riding, judo, wrestling, hurdling, soccer (side passes), swimming (breaststroke legs), general maneuvering on court (i.e. crossover steps, side shifting).

### Movements or injuries that may damage this muscle
Side splits or high side kicks without sufficient warm-up.

### Common problems when muscle is chronically tight/ shortened (spastic)
Groin pulls. (The adductors tend to be much tighter in men than in women.)

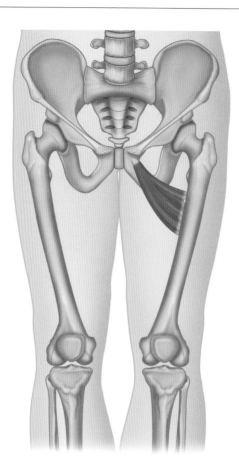

*Anterior view.*

**Latin**, *adducere*, to lead to; *brevis*, short.

Adductor brevis lies anterior to adductor magnus.

**Origin**
Outer surface of inferior ramus of pubis.

**Insertion**
Lower two-thirds of pectineal line and upper half of linea aspera.

**Action**
Adducts hip joint. Flexes extended femur at hip joint. Extends flexed femur at hip joint. Assists in lateral rotation of hip joint.

**Nerve**
Anterior division of obturator nerve (L2–4). Sometimes the posterior division also supplies a branch to it.

**Basic functional movement**
Example: bringing the second leg in or out of a car.

**Sports that heavily utilize this muscle**
Examples: horse riding, judo, wrestling, hurdling, soccer (side passes), swimming (breaststroke legs), general maneuvering on court (i.e. crossover steps, side shifting).

**Movements or injuries that may damage this muscle**
Side splits or high side kicks without sufficient warm-up.

**Common problems when muscle is chronically tight/ shortened (spastic)**
Groin pulls. (The adductors tend to be much tighter in men than in women.)

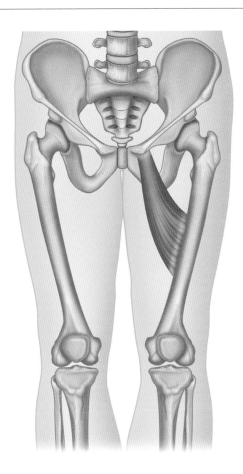

*Anterior view.*

**Latin**, *adducere*, to lead to; *longus*, long.

Adductor longus is the most anterior of the three adductor muscles. The lateral border of its upper fibers form the medial border of the femoral triangle (sartorius forms the lateral boundary, and the inguinal ligament forms the superior boundary).

## Origin
Anterior surface of pubis at junction of crest and symphysis.

## Insertion
Middle third of medial lip of linea aspera.

## Action
Adducts hip joint. Flexes extended femur at hip joint. Extends flexed femur at hip joint. Assists in lateral rotation of hip joint.

## Nerve
Anterior division of obturator nerve L**2**, **3**, 4.

## Basic functional movement
Example: bringing the second leg in or out of a car.

## Sports that heavily utilize this muscle
Examples: horse riding, judo, wrestling, hurdling, soccer (side passes), swimming (breaststroke legs), general maneuvering on court (i.e. crossover steps, side shifting).

## Movements or injuries that may damage this muscle
Side splits or high side kicks without sufficient warm-up.

## Common problems when muscle is chronically tight/ shortened (spastic)
Groin pulls. (The adductors tend to be much tighter in men than in women.)

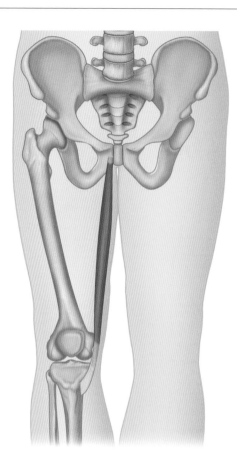

*Anterior view.*

**Latin**, *gracilis*, slender, delicate.

Gracilis descends down the medial side of the thigh, anterior to semimembranosus.

## Origin
Lower half of symphysis pubis and inferior ramus of pubis.

## Insertion
Upper part of medial surface of shaft of tibia.

## Action
Adducts hip joint. Flexes knee joint. Medially rotates knee joint when flexed.

## Nerve
Anterior division of obturator nerve L**2**, **3**, **4**.

## Basic functional movement
Example: sitting with the knees pressed together.

## Sports that heavily utilize this muscle
Examples: horse riding, hurdling, soccer.

## Movements or injuries that may damage this muscle
Side splits or high side kicks without sufficient warm-up.

## Common problems when muscle is chronically tight/shortened (spastic)
Groin pulls. (The adductors tend to be much tighter in men than in women.)

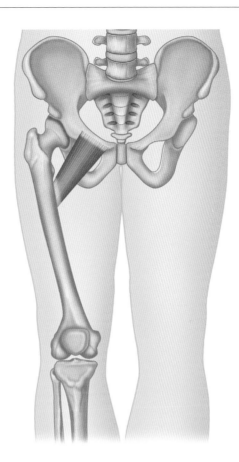

*Anterior view.*

**Latin**, *pecten*, comb; *pectinatus*, comb shaped.

Pectineus is sandwiched between psoas major and adductor longus.

## Origin
Pecten of pubis, between iliopubic (iliopectineal) eminence and pubic tubercle.

## Insertion
Pectineal line, from lesser trochanter to linea aspera of femur.

## Action
Adducts hip joint. Flexes hip joint.

## Nerve
Femoral nerve L**2**, **3**, 4. Occasionally receives an additional branch from obturator nerve L3.

## Basic functional movement
Example: walking along a straight line.

## Sports that heavily utilize this muscle
Examples: horse riding, rugby, sprinting (maximizes stride length), kicking sports (e.g. soccer, to maximize kicking force).

## Movements or injuries that may damage this muscle
Side splits or high side kicks without sufficient warm-up.

## Common problems when muscle is chronically tight/ shortened (spastic)
Groin pulls. (The adductors tend to be much tighter in men than in women.)

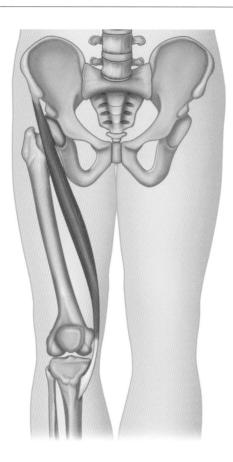

*Anterior view.*

**Latin**, *sartor*, tailor.

Sartorius is the most superficial muscle of the anterior thigh and is also the longest strap muscle in the body. The medial border of the upper third of this muscle forms the lateral boundary of the femoral triangle (adductor longus forms the medial boundary, and the inguinal ligament forms the superior boundary). The action of sartorius is to put the lower limbs in the seated cross-legged position of the tailor (hence its name from the Latin).

## Origin
Anterior superior iliac spine and the area immediately below it.

## Insertion
Upper part of medial surface of tibia, near anterior border.

## Action
Flexes hip joint (helping to bring leg forward in walking or running). Laterally rotates and abducts hip joint. Flexes knee joint. Assists in medial rotation of tibia on femur after flexion. Saying that the muscle makes it possible to place the heel on the knee of the opposite limb may summarize these actions.

## Nerve
Two branches from femoral nerve L2, **3**, (4).

## Basic functional movement
Example: sitting cross-legged.

## Sports that heavily utilize this muscle
Examples: ballet, skating, soccer.

## Movements or injuries that may damage this muscle
Being overambitious with yoga exercises in cross-legged or lotus positions (although the knee is likely to be damaged first).

## Common problems when muscle is chronically tight/shortened (spastic)
Pain or damage to the inside of the knee.

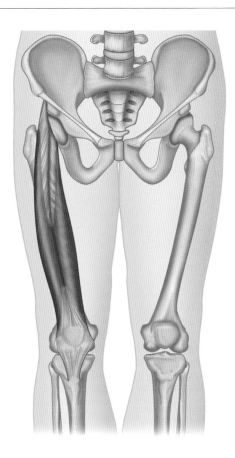

*Anterior view.*

**Latin**, *rectus*, straight; *femoris*, of the thigh.

Rectus femoris is part of the quadriceps femoris, which also includes vastus lateralis, vastus medialis, and vastus intermedius. This spindle-shaped bipennate muscle has two heads of origin: the reflected head is in the line of pull of the muscle in four-footed animals, whereas the straight head seems to have developed in humans as a result of the upright posture.

### Origin
Straight head (anterior head): anterior inferior iliac spine. Reflected head (posterior head): groove above acetabulum (on ilium).

### Insertion
Patella, then via patellar ligament to tuberosity of tibia.

### Action
Extends knee joint and flexes hip joint (particularly in combination, as in kicking a ball). Assists iliopsoas in flexion of trunk on thigh. Prevents flexion at knee joint as heel strikes the ground during walking.

### Nerve
Femoral nerve L**2**, **3**, **4**.

### Basic functional movement
Examples: walking up stairs, cycling.

### Sports that heavily utilize this muscle
Examples: fell running (push-off phase and knee stability when running), skiing, all jumping events, kicking sports (soccer, karate, etc.), weightlifting.

### Common problems when muscle is chronically tight/shortened (spastic)
Low back pain. Knee pain or instability, especially if the muscle is tight and weak.

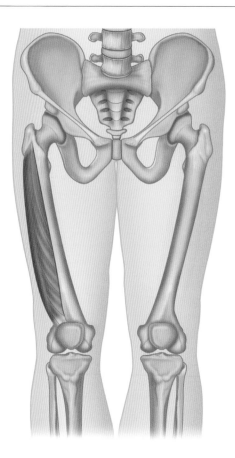

*Anterior view.*

**Latin**, *vastus*, vast; *lateralis*, relating to the side.

Part of the quadriceps femoris. The quadriceps muscles straighten the knee when rising from sitting, during walking, and while climbing. The vasti muscles as a group pay out to control the movement of sitting down.

## Origin
Proximal part of intertrochanteric line. Anterior and inferior borders of greater trochanter. Gluteal tuberosity. Upper half of lateral lip of linea aspera of femur.

## Insertion
Lateral margin of patella, then via patellar ligament to tuberosity of tibia.

## Action
Extends knee joint. Prevents flexion at knee joint as heel strikes the ground during walking.

## Nerve
Femoral nerve L**2**, **3**, **4**.

## Basic functional movement
Examples: walking up stairs, cycling.

## Sports that heavily utilize this muscle
Examples: fell running (push-off phase and knee stability when running), skiing, all jumping events, kicking sports (soccer, karate, etc.), weightlifting.

## Common problems when muscle is chronically tight/ shortened (spastic)
Low back pain. Knee pain or instability, especially if the muscle is tight and weak.

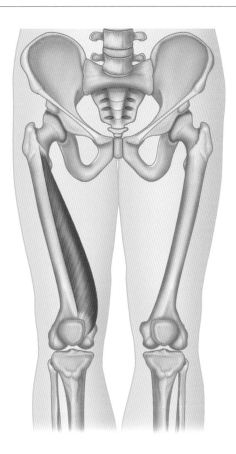

*Anterior view.*

**Latin**, *vastus*, vast; *medialis*, relating to the middle.

Part of the quadriceps femoris. Vastus medialis is larger and heavier than vastus lateralis.

### Origin
Distal one-half of intertrochanteric line. Medial lip of linea aspera. Medial supracondylar line. Medial intermuscular septum.

### Insertion
Medial margin of patella, then via patellar ligament to tuberosity of tibia. Medial condyle of tibia.

### Action
Extends knee joint. Prevents flexion at knee joint as heel strikes the ground during walking.

### Nerve
Femoral nerve L**2**, **3**, **4**.

### Basic functional movement
Examples: walking up stairs, cycling.

### Sports that heavily utilize this muscle
Examples: fell running (push-off phase and knee stability when running), skiing, all jumping events, kicking sports (soccer, karate, etc.), weightlifting.

### Common problems when muscle is chronically tight/shortened (spastic)
Low back pain. Knee pain or instability, especially if the muscle is tight and weak.

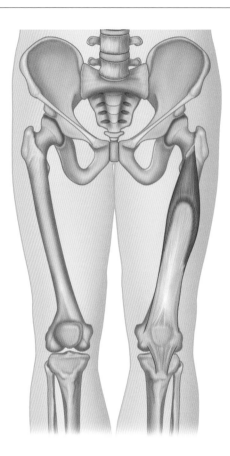

*Anterior view.*

**Latin**, *vastus*, vast; *intermedius*, intermediate.

Vastus intermedius is the deepest part of the quadriceps femoris. This muscle has a membranous tendon on its anterior surface, to allow a gliding movement between itself and the rectus femoris that overlies it.

## Origin

Anterior and lateral surfaces of upper two-thirds of shaft of femur. Lower half of linea aspera. Lateral intermuscular septum. Upper part of lateral supracondylar line.

## Insertion

Deep surface of quadriceps tendon, then via patellar ligament to tuberosity of tibia.

## Action

Extends knee joint. Prevents flexion at knee joint as heel strikes the ground during walking.

## Nerve

Femoral nerve L**2**, **3**, **4**.

## Basic functional movement

Examples: walking up stairs, cycling.

## Sports that heavily utilize this muscle

Examples: fell running (push-off phase and knee stability when running), skiing, all jumping events, kicking sports (soccer, karate, etc.), weightlifting.

## Common problems when muscle is chronically tight/ shortened (spastic)

Low back pain. Knee pain or instability, especially if the muscle is tight and weak.

# 9 Muscles of the Leg and Foot

The muscles of the shin originate at the top of the tibia, just below the knee joint, and extend down the front of the shin and over the ankle joint. The primary action of the shin muscles is to dorsiflex, extend, or invert the ankle joint.

**Tibialis anterior** originates from the lateral condyle of the tibia, and inserts into the medial and plantar surfaces of the medial cuneiform bone. Tibialis anterior is responsible for dorsiflexing and inverting the foot and is used frequently during running to "toe up" with each step. Pain in the front of the shin occurs when the muscle and tendon become inflamed and irritated through overuse or improper form.

The **extensor hallucis longus** and **extensor digitorum longus** are the main extensor muscles of the toes; their tendons run over the front of the ankle and over the foot, and attach to the toes. These muscles dorsiflex the foot and work in opposition to the flexor muscles. When the calf muscles are tight, or the muscles are worked beyond their exertion level, inflammation of the tendons may occur.

**Fibularis (peroneus) tertius, fibularis (peroneus) longus**, and **fibularis (peroneus) brevis** form the lateral compartment of the lower leg. All these muscles pronate, but the last two also act as plantar flexors and evertors at the ankle joint, as well as preventers of inversion and protectors against ankle sprain. The course of the tendon of the insertion of fibularis (peroneus) longus helps maintain the transverse and lateral longitudinal arches of the foot.

The calf muscles are located on the posterior (rear) of the leg and lower portion of the femur. They attach from the heel bone (calcaneus) to their origin on the medial and lateral condyles of the femur, just above the knee joint. The primary actions of the calf muscles are to plantar flex the ankle joint and flex the knee.

The superficial **gastrocnemius** has two heads and crosses two joints: the knee and the ankle. It is part of the composite muscle known as *triceps surae*, which also includes soleus and plantaris. Triceps surae forms the prominent contour of the calf. Gastrocnemius is quite a thin muscle when compared with the thick soleus. As well as plantar flexing the ankle, gastrocnemius assists in flexion of the knee joint, and is a main propelling force in walking and running. Explosive sprinting, for example, may rupture the Achilles tendon at its junction with the muscle belly of gastrocnemius, hence the need to keep the muscle well stretched.

**Plantaris**, a small muscle, is a weak plantar flexor of the ankle, but plays an important neurological role in assessing and adjusting the tension in the Achilles tendon. The long slender tendon of plantaris (the longest tendon in the body) is equivalent to the tendon of palmaris longus in the arm. Interestingly, plantaris is thought to be what remains of a larger plantar flexor of the foot.

Part of triceps surae, **soleus** is so called because its shape resembles a fish. This is a good example of how anatomical nomenclature does little to inform a reader regarding the possible function of a named structure; in this case, something has been named simply because it looks like a fish. Soleus is deep to gastrocnemius, but its medial and lateral fibers bulge from the sides of the leg and extend further distally than gastrocnemius. As well as helping to plantar flex the ankle, soleus assists in flexing the knee. The constant wearing of high-heeled shoes tends to cause this muscle to shorten, which can affect postural integrity.

**Popliteus** is a thin, flat, triangular muscle forming the floor of the distal portion of the popliteal fossa located on the posterior surface of the knee. It is continuous above with the femur by means of a strong tendinous attachment to the lateral femoral condyle. Interestingly, this muscle is also continuous below with the joint capsule, specifically the lateral meniscus, and continues to the head of the fibula via the arcuate popliteal ligament and the medial two-thirds of the superior tibia on the soleal line.

**Flexor digitorum longus, flexor hallucis longus,** and **tibialis posterior** form the deep posterior compartment of the lower leg. Tibialis posterior is the deepest muscle and helps maintain the arches of the foot. Flexor hallucis longus helps maintain the medial longitudinal arch of the foot, while flexor digitorum longus flexes the phalanges of the second to fifth toes and plantar flexes and inverts the ankle.

The feet and ankles are comprised of a multitude of small muscles that control the foot. The muscles around this region, along with the structure of the joints, allow a large range of motion of the feet and ankles, including flexion, extension, adduction, abduction, and rotation.

There are four layers of muscles in the sole of the foot. The first layer is the most inferior (i.e. the most superficial and closest to the ground in standing), comprising **abductor hallucis, flexor digitorum brevis,** and **abductor digiti minimi**. Abductor digiti minimi forms the lateral margin of the sole of the foot. The second layer contains the **lumbricales** and **quadratus plantae**, along with the tendons of flexor hallucis longus and flexor digitorum longus. The third layer contains **flexor hallucis brevis, adductor hallucis,** and **flexor digiti minimi brevis**. The fourth layer is the deepest (and most superior) layer of muscles of the sole of the foot; it consists of the four muscles of the **dorsal interossei** and the three muscles of the **plantar interossei**, as well as the tendons of tibialis posterior and peroneus longus. On the dorsum of the foot lies **extensor digitorum brevis**.

One structure worth noting is a tough fibrous tissue connecting the heel to the toes—the **plantar fascia**, also called the *plantar aponeurosis*. Repetitive ankle movement, especially when restricted by tight calves, can irritate this tissue at the insertion on the heel. Specific dynamic stretches may help to alleviate this problem.

# Muscles of the Leg

The leg comprises three muscle groups: 1) the extensors (dorsiflexors), within the anterior compartment; 2) the peroneal compartment on the lateral side; and 3) the flexors (plantar flexors), within the posterior compartment.

**Strengthen**

Seated calf raise

Squat jumps

Standing calf raise

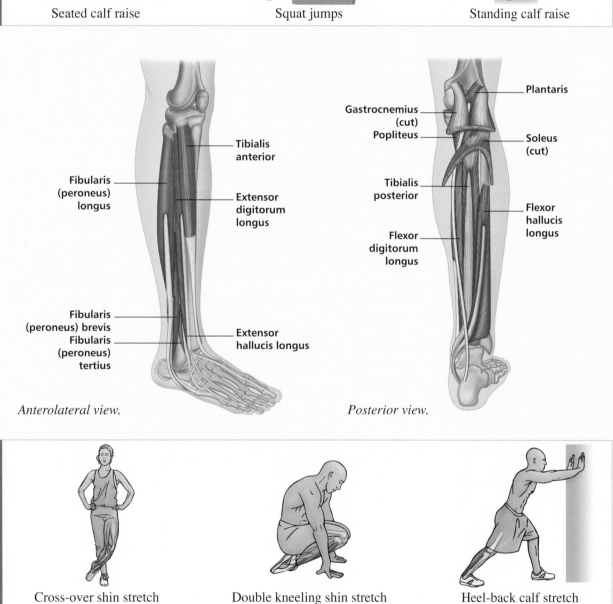

Tibialis anterior

Fibularis (peroneus) longus

Extensor digitorum longus

Fibularis (peroneus) brevis

Fibularis (peroneus) tertius

Extensor hallucis longus

*Anterolateral view.*

Plantaris

Gastrocnemius (cut)

Popliteus

Soleus (cut)

Tibialis posterior

Flexor hallucis longus

Flexor digitorum longus

*Posterior view.*

**Self-stretch**

Cross-over shin stretch

Double kneeling shin stretch

Heel-back calf stretch

*Anterolateral view, right leg.*

**Latin**, *tibialis*, relating to the shin; *anterior*, at the front.

**Origin**
Lateral condyle of tibia. Upper half of lateral surface of tibia. Interosseous membrane.

**Insertion**
Medial and plantar surface of medial cuneiform bone. Base of first metatarsal.

**Action**
Dorsiflexes ankle joint. Inverts ankle joint.

**Nerve**
Deep fibular (peroneal) nerve L4, **5**, S1.

**Basic functional movement**
Example: walking and running (helps prevent the foot from slapping onto the ground after the heel strikes, and lifts the foot clear of the ground as the leg swings forward).

**Sports that heavily utilize this muscle**
Examples: hill walking, mountaineering, running, breaststroke swimming, cycling (the pedal up phase).

**Movements or injuries that may damage this muscle**
Excessive jumping onto hard surfaces.

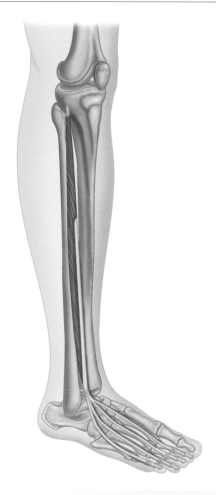

*Anterolateral view, right leg.*

**Latin**, *extendere*, to extend; *digitorum*, of the toes/fingers; *longus*, long.

Like the corresponding tendons in the hand, this muscle forms extensor hoods on the dorsum of the proximal phalanges of the foot. These hoods are joined by the tendons of the lumbricales and extensor digitorum brevis, but not by the interossei.

### Origin
Lateral condyle of tibia. Upper two-thirds of anterior surface of fibula. Upper part of interosseous membrane.

### Insertion
Along dorsal surface of the four lateral toes. Each tendon divides, to attach to bases of middle and distal phalanges.

### Action
Extends toes at metatarsophalangeal joints. Assists in extension of interphalangeal joints. Assists in dorsiflexion and eversion of ankle joint.

### Nerve
Deep fibular (peroneal) nerve L**4, 5**, S**1**.

### Basic functional movement
Example: walking up stairs (ensuring the toes clear the steps).

### Sports that heavily utilize this muscle
Examples: hill walking, mountaineering, running, breaststroke swimming, cycling (the pedal up phase).

### Movements or injuries that may damage this muscle
Tendon is easily bruised by compression (e.g. if toe is stepped on).

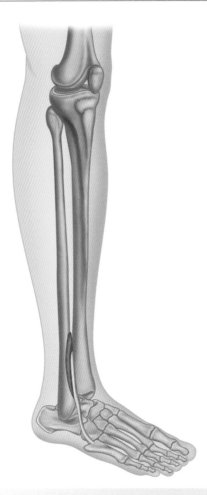

*Anterolateral view, right leg.*

**Latin**, *fibula*, pin/buckle; *tertius*, third. **Greek**, *perone*, pin/buckle.

This muscle is a partially separated, lower lateral part of extensor digitorum longus.

## Origin
Lower third of anterior surface of fibula and interosseous membrane.

## Insertion
Dorsal surface of base of fifth metatarsal.

## Action
Dorsiflexes ankle joint. Everts ankle joint.

## Nerve
Deep fibular (peroneal) nerve L**4**, **5**, S**1**.

## Basic functional movement
Examples: walking and running.

## Sports that heavily utilize this muscle
Examples: running, soccer, jumping.

## Movements or injuries that may damage this muscle
Forced inversion of the ankle (i.e. overstretching the lateral aspect of the ankle) may create chronic problems with ankle joint stability.

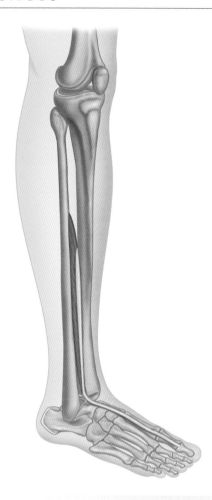

*Anterolateral view, right leg.*

**Latin**, *extendere*, to extend; *hallucis*, of the great toe; *longus*, long.

This muscle lies between, and deep to, tibialis anterior and extensor digitorum longus.

## Origin
Middle half of anterior surface of fibula and adjacent interosseous membrane.

## Insertion
Base of distal phalanx of great toe.

## Action
Extends all joints of great toe. Dorsiflexes ankle joint. Assists in inversion of ankle joint.

## Nerve
Deep fibular (peroneal) nerve L4, 5, S1.

## Basic functional movement
Example: walking up stairs (ensuring the great toe clears the steps).

## Sports that heavily utilize this muscle
Examples: hill walking, mountaineering, breaststroke swimming, cycling (the pedal up phase).

## Movements or injuries that may damage this muscle
Tendon is easily bruised by compression (e.g. if toe is stepped on).

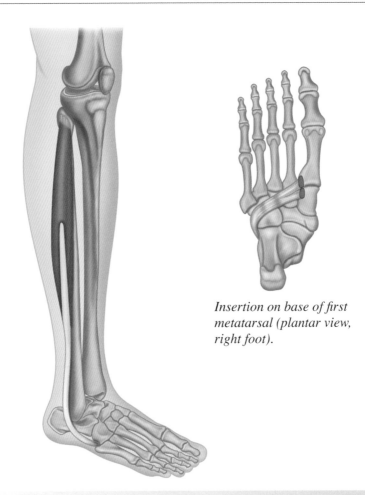

*Insertion on base of first metatarsal (plantar view, right foot).*

*Anterolateral view, right leg.*

**Latin**, *fibula*, pin/buckle; *longus*, long. **Greek**, *perone*, pin/buckle.

The course of the tendon of the insertion of fibularis longus helps maintain the transverse and lateral longitudinal arches of the foot.

**Origin**
Upper two-thirds of lateral surface of fibula. Lateral condyle of tibia.

**Insertion**
Lateral side of medial cuneiform. Base of first metatarsal.

**Action**
Everts ankle joint. Assists in plantar flexion of ankle joint.

**Nerve**
Superficial fibular (peroneal) nerve L4, **5**, S1.

**Basic functional movement**
Example: walking on uneven surfaces.

**Sports that heavily utilize this muscle**
Examples: running, soccer, jumping.

**Movements or injuries that may damage this muscle**
Forced inversion of the ankle (i.e. overstretching the lateral aspect of the ankle) may create chronic problems with ankle joint stability.

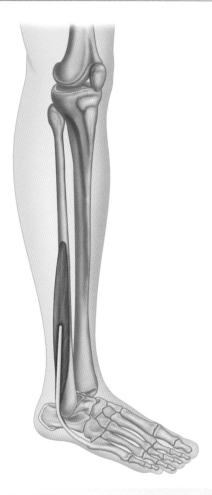

*Anterolateral view, right leg.*

**Latin**, *fibula*, pin/buckle; *brevis*, short. **Greek**, *perone*, pin/buckle.

A slip of muscle from fibularis brevis often joins the long extensor tendon of the little toe, whereupon it is known as *peroneus digiti minimi*.

### Origin
Lower two-thirds of lateral surface of fibula. Adjacent intermuscular septa.

### Insertion
Lateral side of base of fifth metatarsal.

### Action
Everts ankle joint. Assists in plantar flexion of ankle joint.

### Nerve
Superficial fibular (peroneal) nerve L4, **5**, S1.

### Basic functional movement
Example: walking on uneven ground.

### Sports that heavily utilize this muscle
Examples: running, soccer, jumping.

### Movements or injuries that may damage this muscle
Forced inversion of the ankle (i.e. overstretching the lateral aspect of the ankle) may create chronic problems with ankle joint stability.

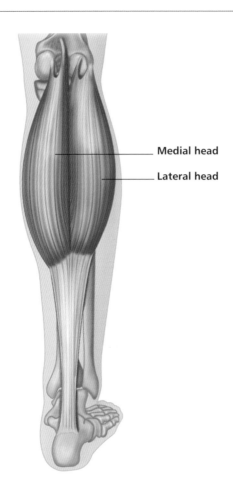

Medial head

Lateral head

*Posterior view, right leg.*

**Greek**, *gaster*, stomach; *kneme*, lower leg.

Gastrocnemius is part of the composite muscle known as *triceps surae*, which also includes soleus and plantaris. Triceps surae forms the prominent contour of the calf. The *popliteal fossa* at the back of the knee is formed inferiorly by the bellies of gastrocnemius and plantaris, laterally by the tendon of biceps femoris, and medially by the tendons of semimembranosus and semitendinosus.

## Origin
Medial head: popliteal surface of femur above medial condyle.
Lateral head: lateral condyle and posterior surface of femur.

## Insertion
Posterior surface of calcaneus (via tendo calcaneus—a fusion of the tendons of gastrocnemius and soleus).

## Action
Plantar flexes foot at ankle joint. Assists in flexion of knee joint. It is a main propelling force in walking and running.

## Nerve
Tibial nerve S1, 2.

## Basic functional movement
Example: standing on tiptoes.

## Sports that heavily utilize this muscle
Examples: most sports requiring running and jumping (especially sprinting, high jump, long jump, volleyball, and basketball), ballet, push-off in the swim start, trampolining.

## Movements or injuries that may damage this muscle
Explosive jumping, or landing badly when jumping down, may rupture the tendo calcaneus (Achilles tendon) at its junction with the muscle belly.

## Common problems when muscle is chronically tight/shortened (spastic)
The constant wearing of high-heeled shoes tends to cause this muscle to shorten, which can affect postural integrity.

# PLANTARIS

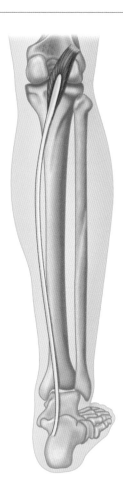

*Posterior view, right leg.*

**Latin**, *plantaris*, relating to the sole.

Part of triceps surae. Its long slender tendon is equivalent to the tendon of palmaris longus in the arm.

**Origin**
Lower part of lateral supracondylar ridge of femur and adjacent part of its popliteal surface. Oblique popliteal ligament of knee joint.

**Insertion**
Posterior surface of calcaneus (or sometimes into the medial surface of tendo calcaneus).

**Action**
Plantar flexes ankle joint. Feebly flexes knee joint.

**Nerve**
Tibial nerve L4, **5**, S1, (2).

**Basic functional movement**
Example: standing on tiptoes.

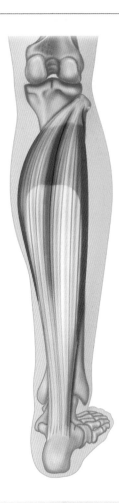

*Posterior view, right leg.*

**Latin**, *solea*, leather sole/sandal/sole (fish).

Soleus is part of triceps surae and is so named because of its shape. The calcaneal tendon of soleus and gastrocnemius is the thickest and strongest tendon in the body.

## Origin
Posterior surfaces of head of fibula and upper third of body of fibula. Soleal line and middle third of medial border of tibia. Tendinous arch between tibia and fibula.

## Insertion
With tendon of gastrocnemius into posterior surface of calcaneus.

## Action
Plantar flexes ankle joint. Soleus is frequently in contraction during standing, to prevent the body falling forward at ankle joint (i.e. to offset the line of pull through the body's center of gravity). Thus, it helps to maintain an upright posture.

## Nerve
Tibial nerve L5, **S1**, **2**.

## Basic functional movement
Example: standing on tiptoes.

## Sports that heavily utilize this muscle
Examples: most sports requiring running and jumping (especially sprinting, high jump, long jump, volleyball, and basketball), ballet, push-off in the swim start, trampolining.

## Movements or injuries that may damage this muscle
Explosive jumping, or landing badly when jumping down, may rupture the tendo calcaneus (Achilles tendon) at its junction with the muscle belly.

## Common problems when muscle is chronically tight/shortened (spastic)
Tight and painful calves or tendo calcaneus (which is usually more a problem of soleus than gastrocnemius). The constant wearing of high-heeled shoes tends to cause this muscle to shorten, which can affect postural integrity.

# POPLITEUS

*Posterior view, right leg.*

**Latin**, *poples*, the ham.

The tendon from the origin of popliteus lies inside the capsule of the knee joint.

## Origin
Lateral surface of lateral condyle of femur. Oblique popliteal ligament of knee joint.

## Insertion
Upper part of posterior surface of tibia, superior to soleal line.

## Action
Laterally rotates femur on tibia when foot is fixed on the ground. Medially rotates tibia on femur when leg is non-weight-bearing. Assists in flexion of knee joint (popliteus "unlocks" the extended knee joint to initiate flexion of leg). Helps reinforce posterior ligaments of knee joint.

## Nerve
Tibial nerve L**4**, **5**, S**1**.

## Basic functional movement
Example: walking.

## Sports that heavily utilize this muscle
All activities involving running and walking.

## Movements or injuries that may damage this muscle
High kicks without sufficient warm-up.

## Common problems when muscle is chronically tight/shortened (spastic)
Inability to fully extend the knee joint, possibly resulting in knee pain or injury.

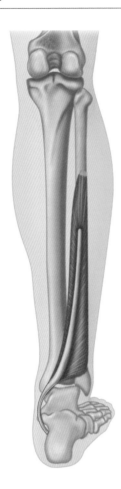

*Posterior view, right leg.*

**Latin**, *flectere*, to bend; *digitorum*, of the toes/fingers; *longus*, long.

The insertion of the tendons of this muscle into the lateral four toes parallels the insertion of flexor digitorum profundus in the hand.

**Origin**
Medial part of posterior surface of tibia, below soleal line.

**Insertion**
Bases of distal phalanges of second through fifth toes.

**Action**
Flexes all joints of lateral four toes (enabling foot to firmly grip the ground when walking). Helps to plantar flex and invert ankle joint.

**Nerve**
Tibial nerve **L5**, **S1**, (2).

**Basic functional movement**
Examples: walking (especially bare foot on uneven ground), standing on tiptoes.

**Sports that heavily utilize this muscle**
Examples: ballet, gymnastics (beam work), karate (side kick).

**Common problems when muscle is chronically tight/shortened (spastic)**
Hammer toe deformity of the lateral four toes.

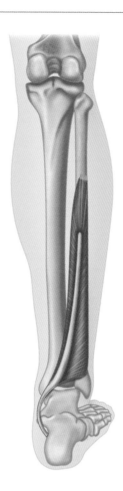

*Posterior view, right leg.*

**Latin**, *flectere*, to bend; *hallucis*, of the great toe; *longus*, long.

This muscle helps maintain the medial longitudinal arch of the foot.

## Origin
Lower two-thirds of posterior surface of fibula. Interosseous membrane. Adjacent intermuscular septum.

## Insertion
Base of distal phalanx of great toe.

## Action
Flexes all joints of great toe, and is important in the final propulsive thrust of foot during walking. Helps to plantar flex and invert ankle joint.

## Nerve
Tibial nerve L5, **S1**, **2**.

## Basic functional movement
Examples: pushing off the surface in walking (especially bare foot on uneven ground), standing on tiptoes.

## Sports that heavily utilize this muscle
Examples: running, hill walking, ballet, gymnastics.

## Common problems when muscle is chronically tight/ shortened (spastic)
Hammer toe deformity of the great toe.

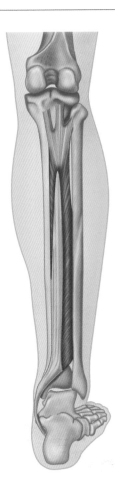

*Posterior view, right leg.*

**Latin**, *tibialis*, relating to the shin; *posterior*, at the back.

Tibialis posterior is the deepest muscle on the back of the leg. It helps maintain the arches of the foot.

## Origin

Lateral part of posterior surface of tibia. Upper two-thirds of posterior surface of fibula. Most of interosseous membrane.

## Insertion

Tuberosity of navicular. By fibrous expansions to sustentaculum tali, three cuneiforms, cuboid, and bases of second, third, and fourth metatarsals.

## Action

Inverts ankle joint. Assists in plantar flexion of ankle joint.

## Nerve

Tibial nerve L(4), **5**, S**1**.

## Basic functional movement

Examples: standing on tiptoes, pushing down car pedals.

## Sports that heavily utilize this muscle

Examples: sprinting, long jump, triple jump.

## Movements or injuries that may damage this muscle

Poor alignment of the lower limb, especially walking or standing with feet turned out, may cause collapse of the medial longitudinal arch of the foot.

# Muscles of the Foot

There are four layers of muscle in the sole of the foot: the first layer is the most inferior (i.e. the most superficial and closest to the ground in standing), and the fourth layer is the deepest (and most superior) layer.

Calf raises      Goosesteps      Single-leg calf raise

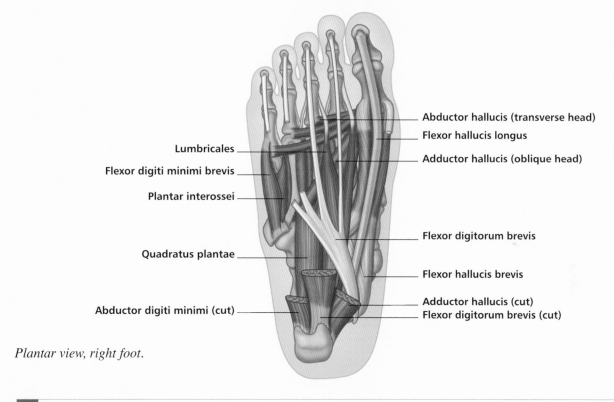

Lumbricales

Flexor digiti minimi brevis

Plantar interossei

Quadratus plantae

Abductor digiti minimi (cut)

Abductor hallucis (transverse head)

Flexor hallucis longus

Adductor hallucis (oblique head)

Flexor digitorum brevis

Flexor hallucis brevis

Adductor hallucis (cut)

Flexor digitorum brevis (cut)

*Plantar view, right foot.*

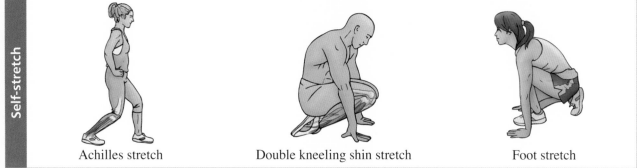

Achilles stretch      Double kneeling shin stretch      Foot stretch

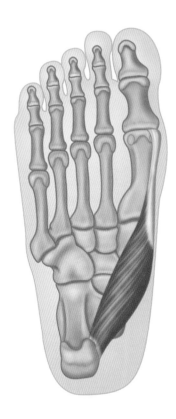

*Plantar view, right foot.*

**Latin**, *abducere*, to lead away from; *hallucis*, of the great toe.

Abductor hallucis forms the medial margin of the sole of the foot.

### Origin
Tuberosity of calcaneus. Flexor retinaculum. Plantar aponeurosis.

### Insertion
Medial side of base of proximal phalanx of great toe.

### Action
Abducts and helps flex great toe at metatarsophalangeal joint.

### Nerve
Medial plantar nerve L4, **5**, S1.

### Basic functional movement
Helps foot stability and power in walking and running.

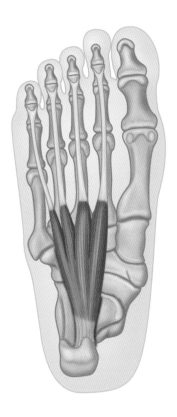

*Plantar view, right foot.*

**Latin**, *flectere*, to bend; *digitorum*, of the toes/fingers; *brevis*, short.

Flexor digitorum brevis is equivalent to the flexor digitorum superficialis muscle of the arm.

**Origin**
Tuberosity of calcaneus. Plantar aponeurosis. Adjacent intermuscular septa.

**Insertion**
Middle phalanges of second to fifth toes.

**Action**
Flexes all joints of lateral four toes, except distal interphalangeal joints.

**Nerve**
Medial plantar nerve L4, **5**, S**1**.

**Basic functional movement**
Helps foot stability and power in walking and running.

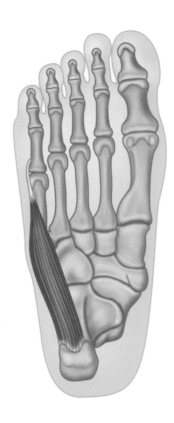

*Plantar view, right foot.*

**Latin**, *abducere*, to lead away from; *digiti*, of the toe/finger; *minimi*, of the smallest.

Abductor digiti minimi forms the lateral margin of the sole of the foot.

**Origin**
Tuberosity of calcaneus. Plantar aponeurosis. Adjacent intermuscular septa.

**Insertion**
Lateral side of base of proximal phalanx of fifth toe.

**Action**
Abducts fifth toe.

**Nerve**
Lateral plantar nerve S**2**, 3.

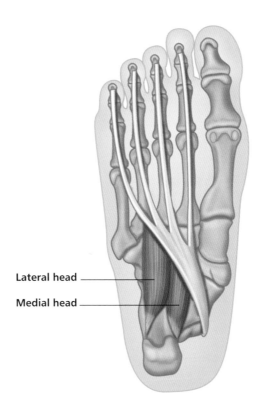

Lateral head

Medial head

*Plantar view, right foot.*

**Latin**, *quadratus*, squared; *plantae*, of the sole.

Quadratus plantae has no counterpart in the hand.

**Origin**
Medial head: medial surface of calcaneus.
Lateral head: lateral border of inferior surface of calcaneus.

**Insertion**
Lateral border of tendon of flexor digitorum longus.

**Action**
Flexes distal phalanges of second to fifth toes. Modifies oblique line of pull of flexor digitorum longus tendons, to bring it in line with long axis of foot.

**Nerve**
Lateral plantar nerve S1, 2.

**Basic functional movement**
Example: holding a pencil between the toes and the ball of the foot.

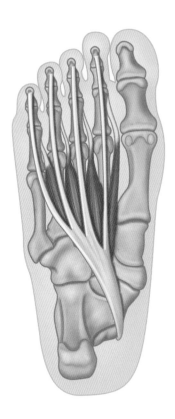

*Plantar view, right foot.*

**Latin**, *lumbricus*, earthworm.

**Origin**
Tendons of flexor digitorum longus.

**Insertion**
Medial side of base of proximal phalanges of second to fifth toes and corresponding extensor expansion.

**Action**
Flex metatarsophalangeal joints and extend interphalangeal joints of lateral four toes.

**Nerve**
Lateral three lumbricales: lateral plantar nerve L(4), (**5**), S**1**, 2.
First lumbricalis: medial plantar nerve L4, **5**, S**1**.

**Basic functional movement**
Example: gathering up material under the foot using the toes only.

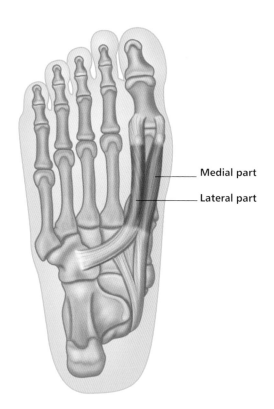

Medial part

Lateral part

*Plantar view, right foot.*

**Latin**, *flectere*, to bend; *hallucis*, of the great toe; *brevis*, short.

The tendons of flexor hallucis brevis contain sesamoid bones. During walking, the great toe pivots on these bones.

### Origin
Medial part of plantar surface of cuboid bone. Adjacent part of lateral cuneiform bone. Tendon of tibialis posterior.

### Insertion
Medial part: medial side of base of proximal phalanx of great toe. Lateral part: lateral side of base of proximal phalanx of great toe.

### Action
Flexes metatarsophalangeal joint of great toe.

### Nerve
Medial plantar nerve L4, **5**, S1.

### Basic functional movement
Example: helping to gather up material under the foot by involving the great toe.

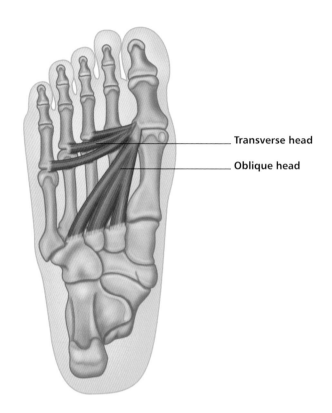

Transverse head

Oblique head

*Plantar view, right foot.*

**Latin**, *adducere*, to lead to; *hallucis*, of the great toe.

Similarly to the adductor of the thumb, adductor hallucis has two heads.

## Origin
Oblique head: bases of second, third, and fourth metatarsals; sheath of peroneus longus tendon. Transverse head: plantar metatarsophalangeal ligaments of third, fourth, and fifth toes; transverse metatarsal ligaments.

## Insertion
Lateral side of base of proximal phalanx of great toe.

## Action
Adducts and assists in flexing metatarsophalangeal joint of great toe.

## Nerve
Lateral plantar nerve S1, 2.

## Basic functional movement
Example: making a space between the great toe and the adjacent toe.

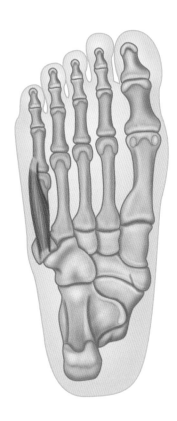

*Plantar view, right foot.*

**Latin**, *flectere*, to bend; *digiti*, of the toe/finger; *minimi*, of the smallest; *brevis*, short.

**Origin**
Sheath of peroneus longus tendon. Base of fifth metatarsal.

**Insertion**
Lateral side of base of proximal phalanx of little toe.

**Action**
Flexes little toe at metatarsophalangeal joint.

**Nerve**
Lateral plantar nerve S2, 3.

**Basic functional movement**
Example: works alongside other toes to gather up material under the foot.

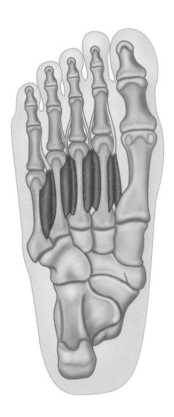

*Dorsal view, left foot.*

**Latin**, *dorsalis*, relating to the back; *interosseus*, between bones.

Similarly to the hand, the dorsal interossei are larger than the plantar interossei.

**Origin**
Adjacent sides of metatarsal bones.

**Insertion**
Bases of proximal phalanges: First: medial side of proximal phalanx of second toe.
Second to fourth: lateral sides of proximal phalanges of second to fourth toes.

**Action**
Abduct (spread) toes. Flex metatarsophalangeal joints.

**Nerve**
Lateral plantar nerve S1, 2.

**Basic functional movement**
Example: facilitate walking.

**Sports that heavily utilize these muscles**
Running, especially with bare feet.

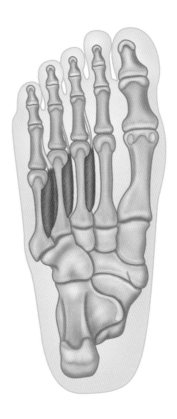

*Plantar view, right foot.*

**Latin**, *plantaris*, relating to the sole; *interosseus*, between bones.

**Origin**
Bases and medial sides of third, fourth, and fifth metatarsals.

**Insertion**
Medial sides of bases of proximal phalanges of same toes.

**Action**
Adduct (close together) toes. Flex metatarsophalangeal joints.

**Nerve**
Lateral plantar nerve S1, 2.

**Basic functional movement**
Example: facilitate walking.

**Sports that heavily utilize these muscles**
Running, especially with bare feet.

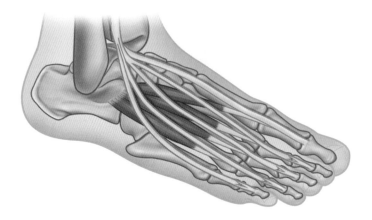

*Anterolateral view, right foot.*

**Latin**, *extendere*, to extend; *digitorum*, of the toes/fingers; *brevis*, short.

This is the only muscle arising from the dorsum of the foot. The part of extensor digitorum brevis that goes to the great toe is called *extensor hallucis brevis*.

**Origin**
Anterior part of superior and lateral surfaces of calcaneus. Lateral talocalcaneal ligament. Inferior extensor retinaculum.

**Insertion**
Base of proximal phalanx of great toe. Lateral sides of tendons of extensor digitorum longus to second, third, and fourth toes.

**Action**
Extends joints of medial four toes.

**Nerve**
Deep fibular (peroneal) nerve L4, **5**, S1.

**Basic functional movement**
Example: facilitates walking.

# Appendix 1: Muscle Innervation Pathways

## Cranial Nerves

Humans are traditionally considered to have twelve pairs of cranial nerves (cranial nerves I–XII), although there are technically thirteen pairs (cranial nerve 0, nervus terminalis). (See "Peripheral Nerve Supply," p. 10). Cranial nerves emerge directly from the brain or brainstem, whereas the spinal nerves emerge directly from the spinal cord. Cranial nerves are listed below, and those that supply the specific skeletal muscles discussed in this book are covered in more detail.

**Cranial nerve 0,** the terminal nerve (nervus terminalis), is also referred to as *cranial nerve XIII*. It may have functional links with the olfactory nerve. **Cranial nerve I**, the olfactory nerve, is responsible for carrying sensory information relating to the sense of smell. **Cranial nerve II**, the optic nerve, is responsible for conveying visual information from the retina to the brain.

**Cranial nerve III,** the oculomotor nerve, controls most of the movements of the eye (along with cranial nerves IV and VI) and innervates the levator palpebrae superioris. **Cranial nerve IV,** the trochlear nerve, is a motor nerve that innervates a single muscle, the superior oblique muscle of the eye (not covered in this book).

**Cranial nerve V,** the trigeminal nerve, is the largest of the cranial nerves and has three main divisions: ophthalmic ($V_1$), maxillary ($V_2$), and mandibular ($V_3$). The trigeminal nerve is responsible for sensation in the face and for functions such as biting and chewing. Both the **ophthalmic division** and the **maxillary division** are purely sensory, while the **mandibular division** has both sensory and motor functions. The mandibular division innervates masseter, temporalis, pterygoids, mylohyoideus, and digastric (anterior belly).

**Cranial nerve VI,** the abducens nerve, controls the movement of just one muscle, the lateral rectus muscle of the eye (not covered in this book).

From the pons of the brain, **cranial nerve VII**, the facial nerve, enters the temporal bone through the internal acoustic meatus, and then emerges through the stylomastoid foramen, where it branches into the **posterior auricular branch**. The five major branches—temporal, zygomatic, buccal, (marginal) mandibular, and cervical (remember the pneumonic "**T**o **Z**anzibar **B**y **M**otor **C**ar")—innervate the facial muscles as follows. **Temporal branches**: frontalis, temporoparietalis, auricularis anterior and superior, orbicularis oculi (also innervated by the zygomatic branches), procerus, and corrugator supercilii. **Zygomatic branches**: orbicularis oculi (also innervated by the temporal branches) and zygomaticus major (also innervated by the buccal branches). **Buccal branches**: depressor septi nasi, orbicularis oris (also innervated by the mandibular branches), levator labii superioris, levator anguli oris, nasalis, zygomaticus major (also innervated by the zygomatic branches), zygomaticus minor, depressor anguli oris, risorius, and buccinator. **Mandibular branches**: orbicularis oris (also innervated by the buccal branches), depressor labii inferioris, depressor anguli oris (also innervated by the buccal branches), mentalis, and stylohyoideus. **Cervical branches**: platysma. Furthermore, the posterior auricular branch subdivides into the **auricular branch**, which innervates the auricularis posterior, and the **occipital branch**, which innervates the occipitalis. The **digastric branch**, which arises close to the stylomastoid foramen, innervates digastric. Certain texts have identified a ramification of the facial nerve, referring to it as the *intermediary nerve*; research has shown this to be a separate and autonomous entity. The intermediary nerve is involved with taste perception, salivation, and lachrymation. If, in time, this nerve were to be regarded as an autonomous structure, we would then be considered to have fourteen (XIV) cranial nerves.

**Cranial nerve VIII,** the vestibulocochlear nerve (also known as the *auditory vestibular nerve*), transmits sound and equilibrium (balance) information from the inner ear to the brain.

**Cranial nerve IX,** the glossopharyngeal nerve, originates from the medulla oblongata and exits the skull through the jugular foramen. Its main function is sensory.

**Cranial nerve X**, the vagus nerve, supplies motor parasympathetic fibers to all organs except the adrenal glands.

**Cranial nerve XI**, the accessory nerve, is unique in that it is formed by both cranial and spinal components that combine and then diverge, with the cranial portion joining the vagus nerve (X), and the spinal portion descending to innervate sternocleidomastoideus and trapezius.

**Cranial nerve XII**, the hypoglossal nerve, innervates muscles of the tongue, although geniohyoideus is innervated by the fibers of cervical nerve C1, conveyed by the hypoglossal nerve X11.

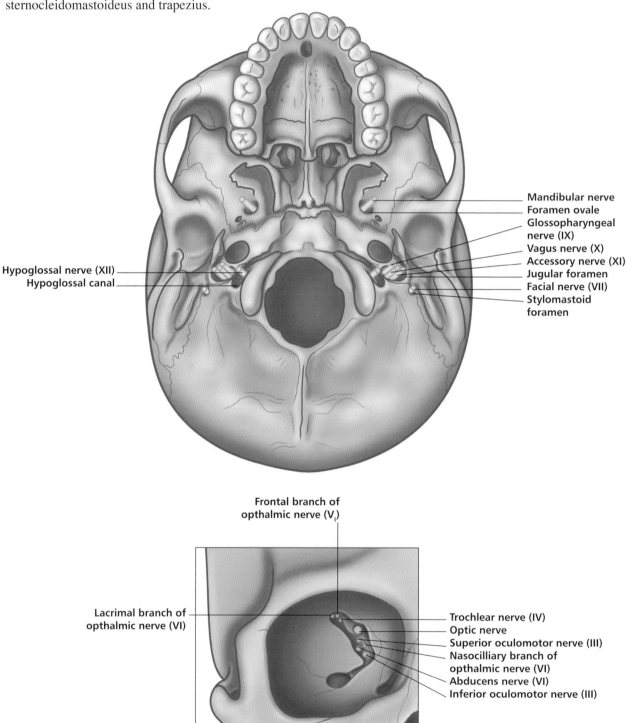

Mandibular nerve
Foramen ovale
Glossopharyngeal nerve (IX)
Vagus nerve (X)
Accessory nerve (XI)
Jugular foramen
Facial nerve (VII)
Stylomastoid foramen

Hypoglossal nerve (XII)
Hypoglossal canal

Frontal branch of opthalmic nerve (V$_I$)

Lacrimal branch of opthalmic nerve (VI)

Trochlear nerve (IV)
Optic nerve
Superior oculomotor nerve (III)
Nasocilliary branch of opthalmic nerve (VI)
Abducens nerve (VI)
Inferior oculomotor nerve (III)

*Right orbit.*

*Cranial Nerves and Skull Passageways (External View).*

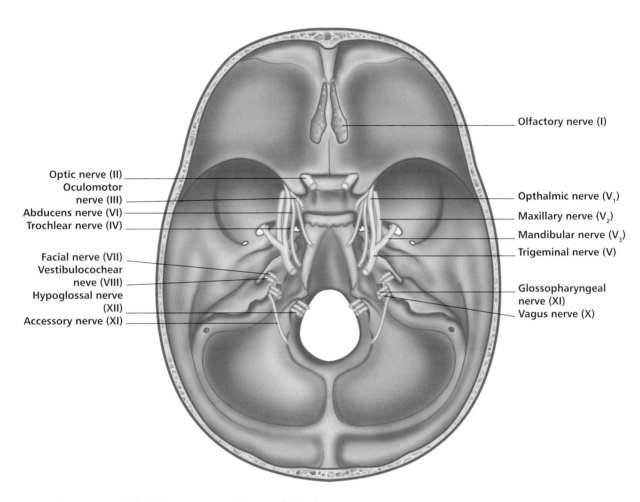

Olfactory nerve (I)

Optic nerve (II)
Oculomotor nerve (III)
Abducens nerve (VI)
Trochlear nerve (IV)

Opthalmic nerve (V$_1$)
Maxillary nerve (V$_2$)
Mandibular nerve (V$_3$)
Trigeminal nerve (V)

Facial nerve (VII)
Vestibulocochear neve (VIII)
Hypoglossal nerve (XII)
Accessory nerve (XI)

Glossopharyngeal nerve (XI)
Vagus nerve (X)

*Cranial Nerves and Skull Passageways (Internal View).*

# Cranial Nerve V—Trigeminal Nerve

## Sensory Distribution

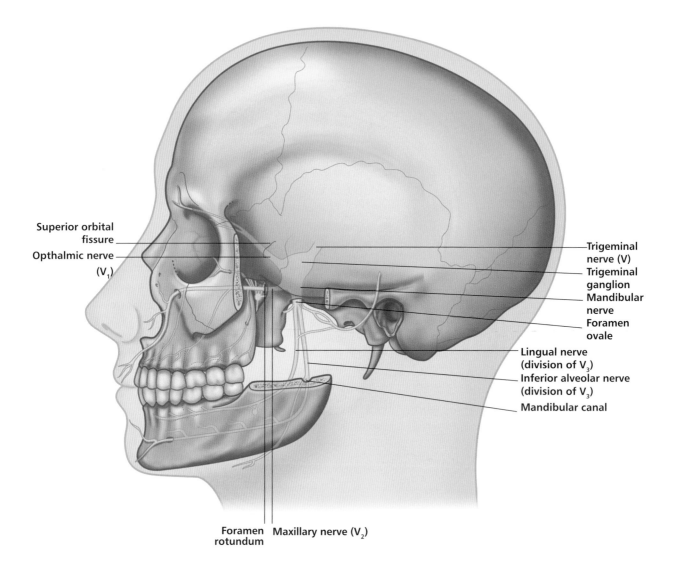

Superior orbital fissure

Opthalmic nerve (V₁)

Trigeminal nerve (V)

Trigeminal ganglion

Mandibular nerve

Foramen ovale

Lingual nerve (division of V₃)

Inferior alveolar nerve (division of V₃)

Mandibular canal

Foramen rotundum

Maxillary nerve (V₂)

# Motor Distribution

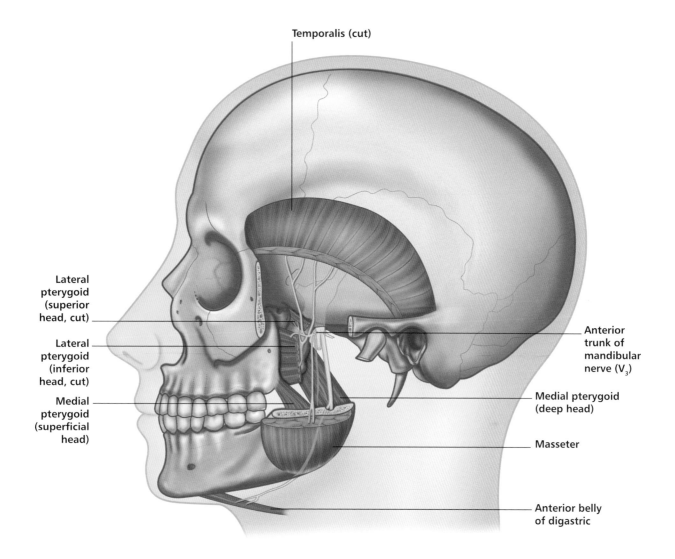

Temporalis (cut)

Lateral pterygoid (superior head, cut)

Lateral pterygoid (inferior head, cut)

Medial pterygoid (superficial head)

Anterior trunk of mandibular nerve (V$_3$)

Medial pterygoid (deep head)

Masseter

Anterior belly of digastric

# Cranial Nerve VII—Facial Nerve

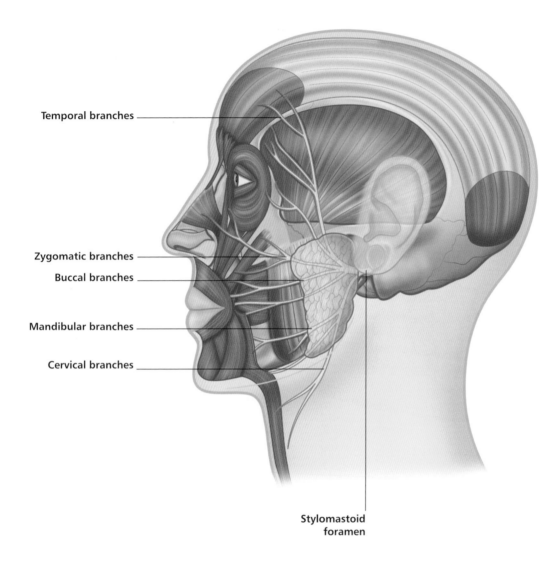

Temporal branches

Zygomatic branches

Buccal branches

Mandibular branches

Cervical branches

Stylomastoid foramen

# Cranial Nerve XI—Accessory Nerve

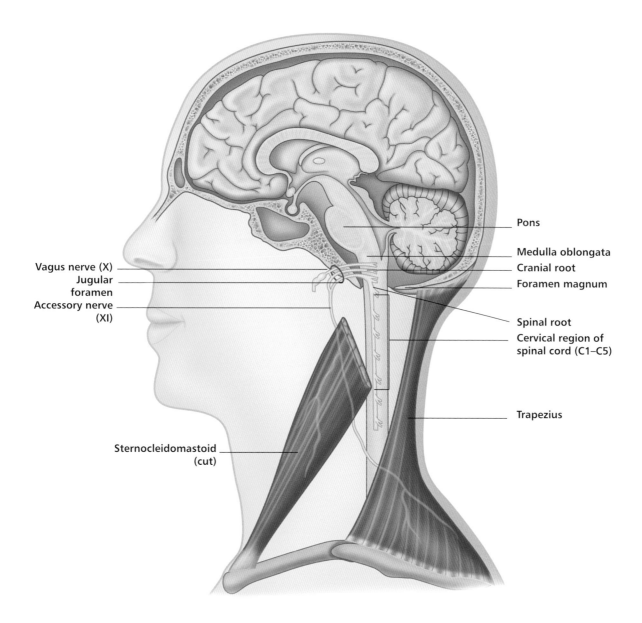

Pons

Medulla oblongata

Cranial root

Foramen magnum

Vagus nerve (X)

Jugular foramen

Accessory nerve (XI)

Spinal root

Cervical region of spinal cord (C1–C5)

Trapezius

Sternocleidomastoid (cut)

## Cervical Plexus

The cervical plexus is a network of nerves, formed by the ventral rami of the four upper cervical nerves (C1–4). The cervical plexus is located in the neck, deep to sternocleidomastoideus, and has two types of branch: cutaneous and muscular. The **muscular branch** comprises: the **ansa cervicalis nerve**, which innervates sternohyoideus, sternothyroideus, thyrohyoideus, and omohyoideus; the **phrenic nerve**, which innervates the diaphragm; and **segmental nerves**, which innervate the middle and anterior scaleni. Furthermore, longus colli, longus capitis, rectus capitis lateralis, rectus capitis anterior, rectus capitis posterior major, and rectus capitis posterior minor are also supplied via the cervical plexus. The **medial brachial cutaneous nerve** innervates the skin on the medial brachial side of the arm.

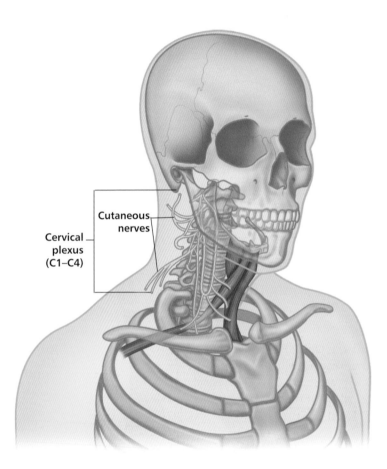

Cutaneous nerves

Cervical plexus (C1–C4)

# Brachial Plexus and Axillary (Circumflex) Nerve

The **brachial plexus** is a network of nerves, formed by the anterior rami of the four lower cervical nerves (C5–8) and first thoracic nerve (T1). The brachial plexus is divided into roots (anterior rami of C5–8 and T1), trunks (superior, middle, inferior), divisions (each of the three trunks splitting in two, to create six divisions), cords (the six divisions regroup to form three cords—lateral, posterior, medial), and finally branches (nerves). Scalenus posterior, rhomboids, latissimus dorsi, supraspinatus, infraspinatus, subscapularis, teres major, and levator scapulae are innervated by the brachial plexus. The five main nerves originating from the brachial plexus are the axillary (circumflex), median, musculocutaneous, ulnar, and radial nerves.

### Axillary (Circumflex) Nerve
The **axillary (circumflex) nerve** carries nerve fibers from C5 and C6, innervating the deltoid and teres minor.

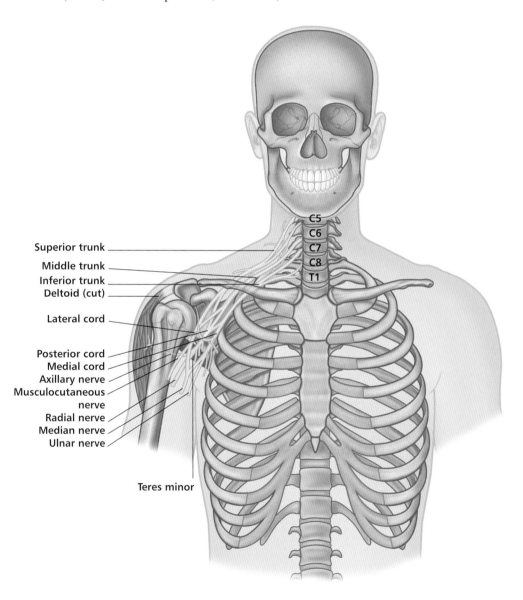

## Musculocutaneous Nerve

The fibers of the **musculocutaneous nerve** are derived from C5–7. It innervates coracobrachialis, biceps brachii, and brachialis. The musculocutaneous nerve is implicated when the patient presents with weak flexion and supination of the forearm. The nerve can be damaged during a bone break at the surgical neck of the humerus, or because of a dislocation. Wearing a heavy bag over one shoulder or carrying a backpack can irritate the nerve.

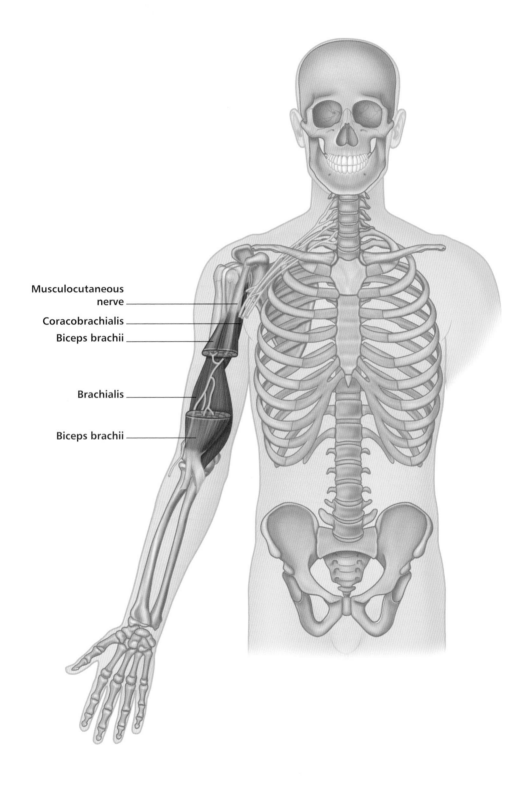

Musculocutaneous nerve

Coracobrachialis

Biceps brachii

Brachialis

Biceps brachii

# Median Nerve

The **median nerve** is the only nerve that supplies the carpal tunnel. It innervates all of the flexors in the forearm, except flexor carpi ulnaris and the medial half of flexor digitorum profundus (both supplied by the ulnar nerve). In the forearm, the median nerve innervates pronator teres, flexor carpi radialis, palmaris longus, flexor digitorum superficialis, flexor digitorum profundus (lateral half), flexor pollicis longus, and pronator quadratus. In the hand, the nerve innervates flexor pollicis brevis (superficial head), opponens pollicis, abductor pollicis brevis, and the first and second lumbricales. The median nerve is implicated when the patient has pain or changes in sensations (such as tingling and numbness) in the first three digits and thumb, or presents with flexor tendonitis.

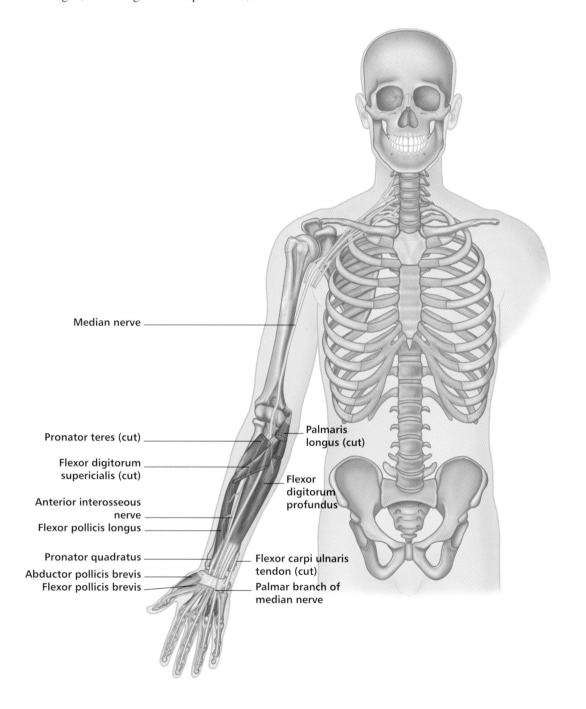

Median nerve

Pronator teres (cut)

Flexor digitorum supericialis (cut)

Anterior interosseous nerve

Flexor pollicis longus

Pronator quadratus

Abductor pollicis brevis

Flexor pollicis brevis

Palmaris longus (cut)

Flexor digitorum profundus

Flexor carpi ulnaris tendon (cut)

Palmar branch of median nerve

# Ulnar Nerve

The **ulnar nerve** originates from the nerve roots of C8–T1 of the brachial plexus. It is the longest unprotected (by either bone or muscle) nerve in the human body and is therefore prone to injury. In the forearm, it subdivides into muscular, palmar, and dorsal branches; in the hand, it further subdivides into superficial and deep branches. The nerve innervates flexor carpi ulnaris, flexor digitorum profundus, adductor pollicis, flexor pollicis brevis (deep head), palmar interossei, abductor digiti minimi, flexor digiti minimi brevis, opponens digiti minimi, palmaris brevis, dorsal interossei, and the third and fourth lumbricales. The ulnar nerve is implicated when the patient complains of pain and changes in sensations (including numbness or tingling in the last two digits), or is experiencing medial epicondylitis, also known as *golfer's elbow.*

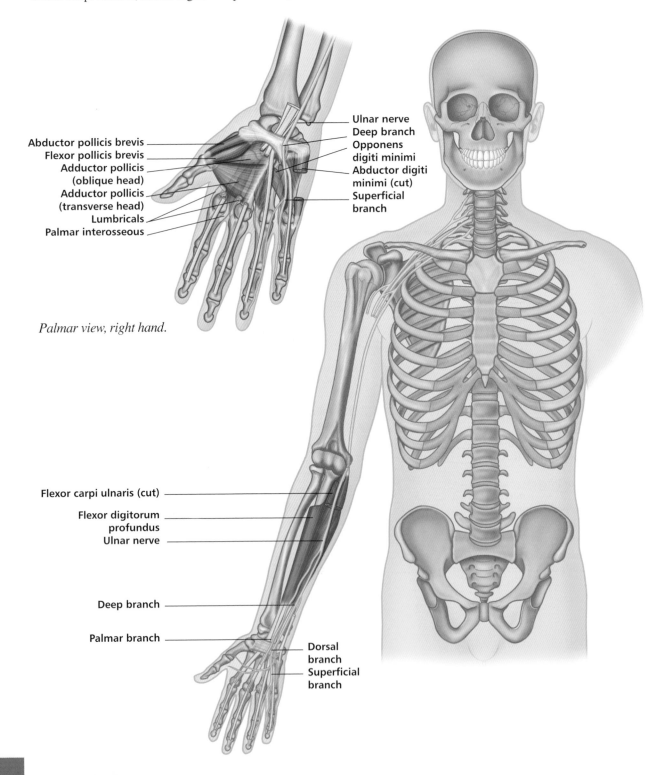

Palmar view, right hand.

Abductor pollicis brevis
Flexor pollicis brevis
Adductor pollicis (oblique head)
Adductor pollicis (transverse head)
Lumbricals
Palmar interosseous

Ulnar nerve
Deep branch
Opponens digiti minimi
Abductor digiti minimi (cut)
Superficial branch

Flexor carpi ulnaris (cut)
Flexor digitorum profundus
Ulnar nerve

Deep branch

Palmar branch

Dorsal branch
Superficial branch

# Radial Nerve

The fibers of the **radial nerve** are derived from C5–T1; the nerve subdivides into muscular and deep branches. The **muscular branches** innervate triceps brachii, anconeus, brachioradialis, and extensor carpi radialis longus. The **deep branch** innervates extensor carpi radialis brevis and supinator. The **posterior interosseous nerve** (a continuation of the deep branch) innervates extensor digitorum, extensor digiti minimi, extensor carpi ulnaris, abductor pollicis longus, extensor pollicis brevis, extensor pollicis longus, and extensor indicis. The radial nerve is implicated when the patient complains of pain and changes in sensations (including numbness or tingling in the posterior proximal half of the first two digits and thumb), or is experiencing lateral epicondylitis, also known as *tennis elbow*.

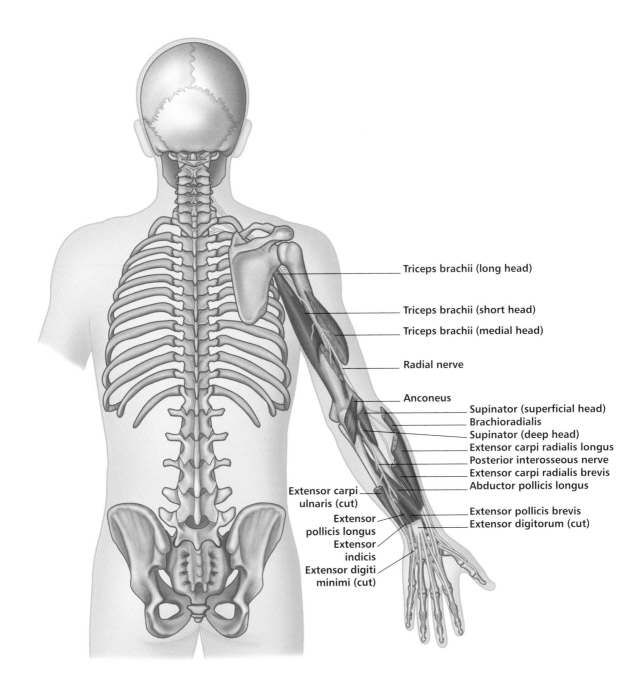

Triceps brachii (long head)

Triceps brachii (short head)

Triceps brachii (medial head)

Radial nerve

Anconeus

Supinator (superficial head)

Brachioradialis

Supinator (deep head)

Extensor carpi radialis longus

Posterior interosseous nerve

Extensor carpi radialis brevis

Abductor pollicis longus

Extensor carpi ulnaris (cut)

Extensor pollicis brevis

Extensor digitorum (cut)

Extensor pollicis longus

Extensor indicis

Extensor digiti minimi (cut)

# Lumbar Plexus

The **lumbar plexus** forms part of the lumbosacral plexus, and is formed by the divisions of the first four lumbar nerves (L1–4) and the subcostal nerve (T12). Branches include: the **ilioinguinal** and **iliohypogastric nerves**, which innervate obliquus internus abdominis and transversus abdominis; the **genitofemoral nerve**, which innervates cremaster; the **inferior gluteal nerve**, which innervates gluteus maximus; and the **superior gluteal nerve**, which innervates tensor fasciae latae, gluteus medius, and gluteus minimus, Also supplied via the lumbosacral plexus are piriformis (nerve to piriformis L5, S1), obturator internus (nerve to obturator L5, S1, 2), gemellus superior and inferior (nerve to obturator L5, S1, 2), and quadratus femoris (nerve to quadratus femoris L4–5). See also the obturator, femoral, sciatic, tibial, and common fibular nerves discussed below.

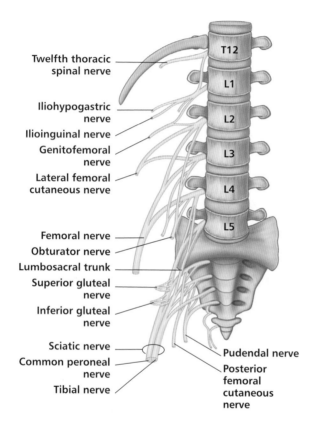

Twelfth thoracic spinal nerve

Iliohypogastric nerve

Ilioinguinal nerve

Genitofemoral nerve

Lateral femoral cutaneous nerve

Femoral nerve

Obturator nerve

Lumbosacral trunk

Superior gluteal nerve

Inferior gluteal nerve

Sciatic nerve

Common peroneal nerve

Tibial nerve

T12

L1

L2

L3

L4

L5

Pudendal nerve

Posterior femoral cutaneous nerve

# Obturator Nerve

The **obturator nerve** origniates from the ventral divisions of the second, third, and fourth lumbar nerves in the lumbar plexus and innervates obturator externus, adductor brevis, adductor magnus, adductor longus, gracilis, and pectineus (occasionally). Despite its name, the obturator nerve is not responsible for the innervation of obturator internus.

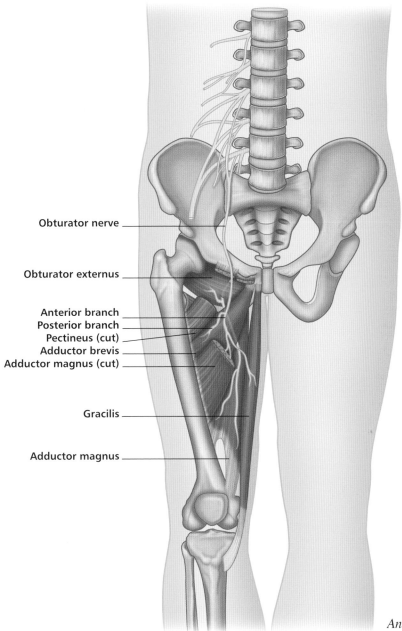

Obturator nerve

Obturator externus

Anterior branch
Posterior branch
Pectineus (cut)
Adductor brevis
Adductor magnus (cut)

Gracilis

Adductor magnus

*Anterior view.*

# Femoral Nerve

The **femoral nerve**, the largest branch of the lumbosacral plexus, is located in the thigh and not in the leg as some texts claim. It originates from the dorsal divisions of the ventral rami of the second, third, and fourth lumbar nerves (L2–4). In the femoral region, the nerve subdivides into the anterior and posterior divisions, before subdividing further into many smaller branches throughout the anterior and medial thigh. The **anterior division** innervates iliacus, sartorius, and pectineus, while the **posterior division** innervates rectus femoris, vastus lateralis, vastus medialis, and vastus intermedius.

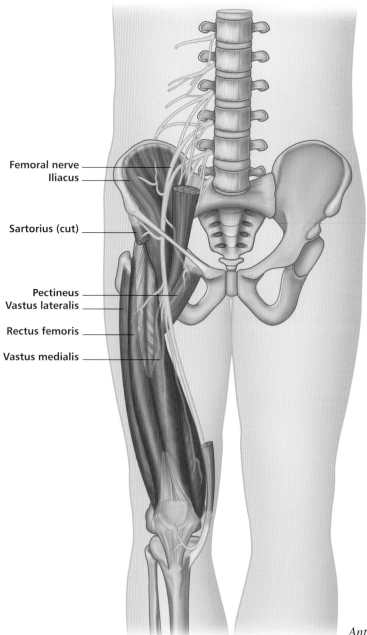

Femoral nerve
Iliacus

Sartorius (cut)

Pectineus
Vastus lateralis

Rectus femoris

Vastus medialis

*Anterior view.*

# Sciatic Nerve

The sciatic nerve is the longest and widest nerve in the human body. It originates in the lower back, from spinal nerves L4 through S3, and passes deep to piriformis down to the lower limb. The sciatic nerve innervates biceps femoris, semimembranosus, and semitendinosus. True sciatic insults include changes in sensations numbness, weakness, and even the sensation of water running down the limb. Depending on the source and level of irritation, the pain can be mild to severe. Sciatic nerve irritation usually occurs at the L5 or S1 level of the spine and only on one side. Pain can travel all the way to the foot and can retard normal motion, but with normal healing, the referred pain should dissipate and become more central. Unresolved chronic pain, especially of unknown origin, should be brought to the attention of the doctor or primary healthcare team.

Midway between the pelvis and the popliteal fossa, the sciatic nerve divides into the tibial nerve and the common fibular (peroneal) nerve.

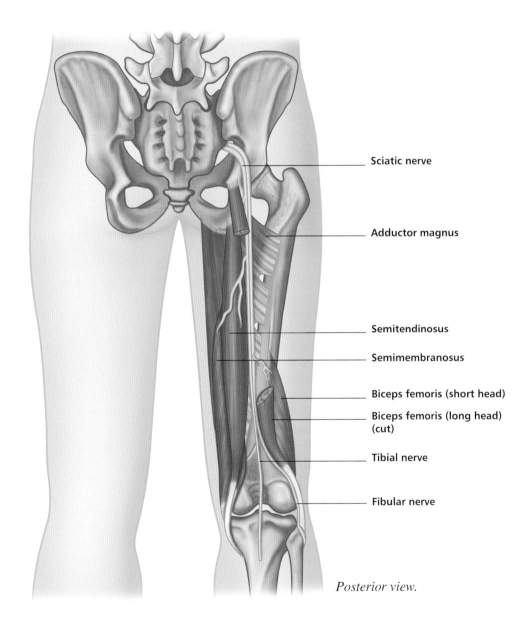

Sciatic nerve

Adductor magnus

Semitendinosus

Semimembranosus

Biceps femoris (short head)

Biceps femoris (long head) (cut)

Tibial nerve

Fibular nerve

*Posterior view.*

# Tibial Nerve

The **tibial nerve** is a branch of the sciatic nerve, and innervates the muscles of the posterior compartment of the leg, including gastrocnemius, plantaris, soleus, flexor digitorum longus, tibialis posterior, popliteus, and flexor hallucis longus. One of its branches, the **medial plantar nerve**, innervates abductor hallucis, flexor digitorum brevis, flexor hallucis brevis, and the first lumbricalis. The other branch, the **lateral plantar nerve**, innervates abductor digiti minimi, quadratus plantae, adductor hallucis, flexor digiti minimi brevis, plantar interossei, dorsal interossei, and the three lateral lumbricales.

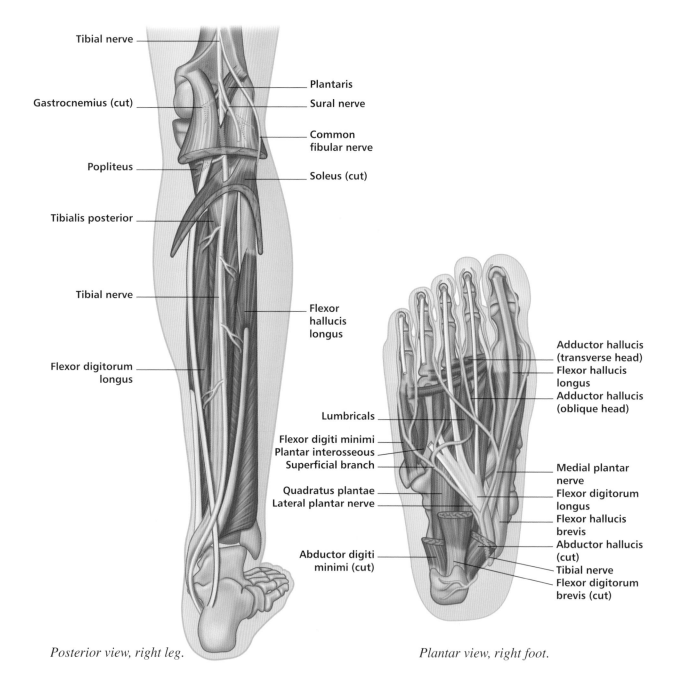

Tibial nerve

Gastrocnemius (cut)

Popliteus

Tibialis posterior

Tibial nerve

Flexor digitorum longus

Plantaris

Sural nerve

Common fibular nerve

Soleus (cut)

Flexor hallucis longus

Lumbricals

Flexor digiti minimi

Plantar interosseous

Superficial branch

Quadratus plantae

Lateral plantar nerve

Abductor digiti minimi (cut)

Adductor hallucis (transverse head)

Flexor hallucis longus

Adductor hallucis (oblique head)

Medial plantar nerve

Flexor digitorum longus

Flexor hallucis brevis

Abductor hallucis (cut)

Tibial nerve

Flexor digitorum brevis (cut)

*Posterior view, right leg.*

*Plantar view, right foot.*

# Common Fibular (Peroneal) Nerve

The **common fibular (peroneal) nerve** originates, via the sciatic nerve, from the dorsal branches of the fourth and fifth lumbar nerves (L4–5) and the first and second sacral nerves (S1–2). It divides into the superficial fibular (peroneal) nerve and the deep fibular (peroneal) nerve. The **superficial fibular (peroneal) nerve** innervates fibularis (peroneus) longus and fibularis (peroneus) brevis. The **deep fibular (peroneal) nerve** innervates tibialis anterior, extensor digitorum longus, fibularis (peroneus) tertius, extensor hallucis longus, extensor hallucis brevis, and extensor digitorum brevis.

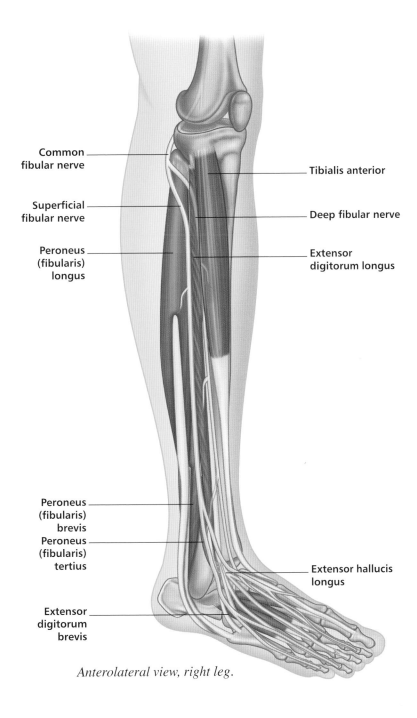

Common fibular nerve

Tibialis anterior

Superficial fibular nerve

Deep fibular nerve

Peroneus (fibularis) longus

Extensor digitorum longus

Peroneus (fibularis) brevis

Peroneus (fibularis) tertius

Extensor hallucis longus

Extensor digitorum brevis

*Anterolateral view, right leg.*

# Appendix 2: Main Muscles Involved in Different Movements of the Body

**MANDIBLE**

*Elevation*
Temporalis (anterior fibers), masseter, pterygoideus medialis

*Depression*
Pterygoideus lateralis, digastricus, mylohyoideus, geniohyoideus

*Protraction*
Pterygoideus lateralis, pterygoideus medialis, masseter (superficial fibers)

*Retraction*
Temporalis (horizontal fibers), digastricus

*Chewing*
Pterygoideus lateralis, pterygoideus medialis, masseter, temporalis


**LARYNX**

*Elevation*
Digastricus, stylohyoideus, mylohyoideus, geniohyoideus, thyrohyoideus

*Depression*
Sternohyoideus, sternothyroideus, omohyoideus

*Protraction*
Geniohyoideus

*Retraction*
Stylohyoideus


**ATLANTO-OCCIPITAL AND ATLANTO-AXIAL JOINTS**

*Flexion*
Longus capitis, rectus capitis anterior, sternocleidomastoideus (anterior fibers)

*Extension*
Semispinalis capitis, splenius capitis, rectus capitis posterior major, rectus capitis posterior minor, obliquus capitis superior, longissimus capitis, trapezius, sternocleidomastoideus (posterior fibers)

*Rotation and Lateral Flexion*
Sternocleidomastoideus, obliquus capitis inferior, obliquus capitis superior, rectus capitis lateralis, longissimus capitis, splenius capitis

## INTERVERTEBRAL JOINTS

### Cervical Region

*Flexion*
Longus colli, longus capitis, sternocleidomastoideus

*Extension*
Longissimus cervicis, longissimus capitis, splenius capitis, splenius cervicis, semispinalis cervicis, semispinalis capitis, trapezius, interspinales, iliocostalis cervicis

*Rotation and Lateral Flexion*
Longissimus cervicis, longissimus capitis, splenius capitis, splenius cervicis, multifidus, longus colli, scalenus anterior, scalenus medius, scalenus posterior, sternocleidomastoideus, levator scapulae, iliocostalis cervicis, intertransversarii

### Thoracic and/or Lumbar Regions

*Flexion*
Muscles of anterior abdominal wall

*Extension*
Erector spinae, quadratus lumborum, trapezius

*Rotation and Lateral Flexion*
Iliocostalis lumborum, iliocostalis thoracis, multifidus, rotatores, intertransversarii, quadratus lumborum, psoas major, muscles of anterior abdominal wall

## SHOULDER GIRDLE

*Elevation*
Trapezius (upper fibers), levator scapulae, rhomboideus minor, rhomboideus major, sternocleidomastoideus

*Depression*
Trapezius (lower fibers), pectoralis minor, pectoralis major (sternocostal portion), latissimus dorsi

*Protraction*
Serratus anterior, pectoralis minor, pectoralis major

*Retraction*
Trapezius (middle fibers), rhomboideus minor, rhomboideus major, latissimus dorsi

*Lateral Displacement of Inferior Angle of Scapula*
Serratus anterior, trapezius (upper and lower fibers)

*Medial Displacement of Inferior Angle of Scapula*
Pectoralis minor, rhomboideus minor, rhomboideus major, latissimus dorsi

## SHOULDER JOINT

*Flexion*
Deltoideus (anterior portion), pectoralis major (clavicular portion; sternocostal portion flexes the extended humerus as far as the position of rest), biceps brachii, coracobrachialis

*Extension*
Deltoideus (posterior portion), teres major (of flexed humerus), latissimus dorsi (of flexed humerus), pectoralis major (sternocostal portion of flexed humerus), triceps brachii (long head to position of rest)

*Abduction*
Deltoideus (middle portion), supraspinatus, biceps brachii (long head)

*Adduction*
Pectoralis major, teres major, latissimus dorsi, triceps brachii (long head), coracobrachialis

*Lateral Rotation*
Deltoideus (posterior portion), infraspinatus, teres minor

*Medial Rotation*
Pectoralis major, teres major, latissimus dorsi, deltoideus (anterior portion), subscapularis

*Horizontal Flexion*
Deltoideus (anterior portion), pectoralis major, subscapularis

*Horizontal Extension*
Deltoideus (posterior portion), infraspinatus

## ELBOW JOINT

*Flexion*
Brachialis, biceps brachii, brachioradialis, extensor carpi radialis longus, pronator teres, flexor carpi radialis

*Extension*
Triceps brachii, anconeus

## RADIOULNAR JOINTS

*Supination*
Supinator, biceps brachii, extensor pollicis longus

*Pronation*
Pronator quadratus, pronator teres, flexor carpi radialis

## RADIOCARPAL AND MIDCARPAL JOINTS

*Flexion*
Flexor carpi radialis, flexor carpi ulnaris, palmaris longus, flexor digitorum superficialis, flexor digitorum profundus, flexor pollicis longus, abductor pollicis longus, extensor pollicis brevis

*Extension*
Extensor carpi radialis brevis, extensor carpi radialis longus, extensor carpi ulnaris, extensor digitorum, extensor indicis, extensor pollicis longus, extensor digiti minimi

*Abduction*
Extensor carpi radialis brevis, extensor carpi radialis longus, flexor carpi radialis, abductor pollicis longus, extensor pollicis longus, extensor pollicis brevis

*Adduction*
Flexor carpi ulnaris, extensor carpi ulnaris

## METACARPOPHALANGEAL JOINTS OF THE FINGERS

*Flexion*
Flexor digitorum profundus, flexor digitorum superficialis, lumbricales, interossei, flexor digiti minimi, abductor digiti minimi, palmaris longus (through palmar aponeurosis)

*Extension*
Extensor digitorum, extensor indicis, extensor digiti minimi

*Abduction and Adduction*
Interossei, abductor digiti minimi, lumbricales (may assist in radial deviation), extensor digitorum (abducts by hyperextending; tendon to index radially deviates), flexor digitorum profundus (adducts by flexing), flexor digitorum superficialis (adducts by flexing)

*Rotation*
Lumbricales, interossei (movement is slight except index; only effective when phalanx is flexed), opponens digiti minimi (rotates little finger at carpometacarpal joint)

## INTERPHALANGEAL JOINTS OF THE FINGERS

*Flexion*
Flexor digitorum profundus (both joints), flexor digitorum superficialis (proximal joint only)

*Extension*
Extensor digitorum, extensor digiti minimi, extensor indicis, lumbricales, interossei

## CARPOMETACARPAL JOINT OF THE THUMB

*Flexion*
Flexor pollicis brevis, flexor pollicis longus, opponens pollicis

*Extension*
Extensor pollicis brevis, extensor pollicis longus, abductor pollicis longus

*Abduction*
Abductor pollicis brevis, abductor pollicis longus

*Adduction*
Adductor pollicis, dorsal interossei (first only), extensor pollicis longus (in full extension/abduction), flexor pollicis longus (in full extension/abduction)

*Opposition*
Opponens pollicis, abductor pollicis brevis, flexor pollicis brevis, flexor pollicis longus, adductor pollicis

## METACARPOPHALANGEAL JOINT OF THE THUMB

*Flexion*
Flexor pollicis brevis, flexor pollicis longus, palmar interossei (first only), abductor pollicis brevis

*Extension*
Extensor pollicis brevis, extensor pollicis longus

*Abduction*
Abductor pollicis brevis

*Adduction*
Adductor pollicis, palmar interossei (first only)

## INTERPHALANGEAL JOINT OF THE THUMB

*Flexion*
Flexor pollicis longus

*Extension*
Abductor pollicis brevis, extensor pollicis longus, adductor pollicis, extensor pollicis brevis (occasional insertion)

## HIP JOINT

*Flexion*
Iliopsoas, rectus femoris, tensor fasciae latae, sartorius, adductor brevis, adductor longus, pectineus

*Extension*
Gluteus maximus, semitendinosus, semimembranosus, biceps femoris (long head), adductor magnus (ischial fibers)

*Abduction*
Gluteus medius, gluteus minimus, tensor fasciae latae, obturator internus (in flexion), piriformis (in flexion)

*Adduction*
Adductor magnus, adductor brevis, adductor longus, pectineus, gracilis, gluteus maximus (lower fibers), quadratus femoris

*Lateral Rotation*
Gluteus maximus, obturator internus, gemelli, obturator externus, quadratus femoris, piriformis, sartorius, adductor magnus, adductor brevis, adductor longus

*Medial Rotation*
Iliopsoas (in initial stage of flexion), tensor fasciae latae, gluteus medius (anterior fibers), gluteus minimus (anterior fibers)

## KNEE JOINT

*Flexion*
Semitendinosus, semimembranosus, biceps femoris, gastrocnemius, plantaris, sartorius, gracilis, popliteus

*Extension*
Quadratus femoris

*Medial Rotation of Tibia on Femur*
Popliteus, semitendinosus, semimembranosus, sartorius, gracilis

*Lateral Rotation of Tibia on Femur*
Biceps femoris

## ANKLE JOINT

*Inversion*
Tibialis anterior, tibialis posterior

*Eversion*
Fibularis (peroneus) longus, fibularis (peroneus) brevis, fibularis (peroneus) tertius.

*Dorsiflexion*
Tibialis anterior, extensor hallucis longus, extensor digitorum longus, fibularis (peroneus) tertius

*Plantar Flexion*
Gastrocnemius, plantaris, soleus, tibialis posterior, flexor hallucis longus, flexor digitorum longus, fibularis (peroneus) longus, fibularis (peroneus) brevis

## INTERTARSAL JOINTS

*Inversion*
Tibialis anterior, tibialis posterior

*Eversion*
Fibularis (peroneus) tertius, fibularis (peroneus) longus, fibularis (peroneus) brevis

*Other Movements*
Sliding movements which allow some dorsiflexion, plantar flexion, abduction, and adduction are produced by the muscles acting on the toes. Tibialis anterior, tibialis posterior, and fibularis (peroneus) tertius are also involved.

## METATARSOPHALANGEAL JOINTS OF THE TOES

*Flexion*
Flexor hallucis brevis, flexor hallucis longus, flexor digitorum longus, flexor digitorum brevis, flexor digiti minimi brevis, lumbricales, interossei

*Extension*
Extensor hallucis longus, extensor digitorum brevis, extensor digitorum longus

*Abduction and Adduction*
Abductor hallucis, adductor hallucis, interossei, abductor digiti minimi

## INTERPHALANGEAL JOINTS OF THE TOES

*Flexion*
Flexor hallucis longus, flexor digitorum brevis (proximal joint only), flexor digitorum longus

*Extension*
Extensor hallucis longus, extensor digitorum brevis (not in great toe), extensor digitorum longus, lumbrial

# Resources

Alter, M.J. 1998. *Sport Stretch: 311 Stretches for 41 Sports*, Champaign, IL: Human Kinetics.

Anderson, D.M. (chief lexicographer) 2003. *Dorland's Illustrated Medical Dictionary*, 30th edn, Philadelphia, PA: Saunders.

Bartelink, D.L. 1957. The role of abdominal pressure in relieving the pressure on the lumbar intervertebral discs. *Journal of Bone and Joint Surgery* 39-B, 718.

Biel, A. 2001. *Trail Guide to the Body*, 2nd edn, Boulder, CO: Books of Discovery.

Bumke, O. and Foerster, O. (eds) 1936. *Handbuch der Neurologie*, Vol. V, Berlin: Julius Springer.

Clemente, C.M. (ed.) 1985. *Gray's Anatomy of the Human Body*, 30th edn, Philadelphia, PA: Lea & Febiger.

DeJong, R.N. 1967. *The Neurological Examination*, 3rd edn, New York: Harper & Row.

Fuller, G.N. and Burger, P.C. 1990. Nervus terminals (cranial nerve zero) in the adult human. Clin. Neuropathol. 9 (6): 279–83.

Gracovetsky, S. 1988. *The Spinal Engine*. New York: Springer-Verlag Wein.

Haymaker, W. and Woodhall, B. 1953. *Peripheral Nerve Injuries,* 2nd edn, Philadelphia, PA: W.B. Saunders Co.

Hodges, P.W. and Richardson, C.A. 1997. Feedforward contraction of transversus abdominis is not influenced by direction of arm movement. Experimental Brain Research 114 (2), 362–370.

Huijing, P.A. and Baan, G.C. 2001. Extramuscular myofascial force transmission within the rat anterior tibial compartment: Proximo-distal differences in muscle force. *Acta Physiologica Scandinavica* 173(3), 297–311.

Huxley, H. and Hanson, J. 1954. *Changes in the cross-striations of muscle during contraction and stretch and their structural interpretation*. Nature 173 (4412), 973–976.

Kendall, F.P. and McCreary, E.K. 1983. *Muscles, Testing & Function*, 3rd edn, Baltimore, MD: Williams & Wilkins.

Lawrence, M. 2004. *Complete Guide to Core Stability*, London: A&C Black.

Levin, S.M. 2002. The tensegrity-truss as a model for spine mechanics. *Journal of Mechanics in Medicine and Biology* 2(3&4), 375–388.

Masi, A.T. and Hannon, J.C. 2008. Human resting muscle tone (HRMT): Narrative introduction and modern concepts. *Journal of Bodywork and Movement Therapies* 12(4), 320–332.

Myers, T.W. 2001. *Anatomy Trains*, Edinburgh: Elsevier.

Norris, C.M. 1997. *Abdominal Training*, London A&C Black.

Romanes, G.J. (ed.) 1972. *Cunningham's Textbook of Anatomy*, 11th edn, London: Oxford University Press.

Scarr, G. 2013. *Biotensegrity: The Structural Basis of Life*, Fountainhall, Scotland: Handspring Publishing.

Schade, J.P. 1966. *The Peripheral Nervous System*, New York: Elsevier.

Sharkey, J. 2014. A new anatomy for the 21st century. *sportEX dynamics* 39, 14–17.

Spalteholz, W. (date unknown). *Hand Atlas of Human Anatomy*, Vols II and III, 6th edn, London: J.B. Lippincott.

Tortora, G. 1989. *Principles of Human Anatomy*, 5th edn, New York: Harper & Row.

# Index

Bursae, 37. *See also* Synovial joints
Buttocks muscles, 219
  gluteus maximus, **217**, **220**
  gluteus medius, **217**, **222**
  gluteus minimus, **217**, **223**
  tensor fasciae latae, **217**, **221**

Calf raises, 261
Capillary plexus, 24. *See also* Skeletal muscles
Capsular ligament, 37
Carpometacarpal joint of thumb, 297. *See also* Body movement muscles
Central nervous system (CNS), 10
Cervical plexus, 280. *See also* Muscle innervation pathways
Cervical region muscles, 294. *See also* Body movement muscles
Circular muscles, 30. *See also* Skeletal muscles
Circumflex nerve, 281. *See also* Muscle innervation pathways
Close-grip bench press, 170
CNS. *See* Central nervous system (CNS)
Collateral or accessory ligaments, 37. *See also* Synovial joints
Common fibular nerve, 291. *See also* Muscle innervation pathways
Concurrent movement, 46. *See also* Musculoskeletal mechanics
Condyloid joints, 39. *See also* Synovial joints
Convergent muscles, 30. *See also* Skeletal muscles
Coracobrachialis, **159**, **179**
Core stability, 48. *See also* Musculoskeletal mechanics
Corrugator supercilii, **53**, **66**. *See also* Eyelid muscles; Face muscles; Scalp muscles
Countercurrent movement, 47. *See also* Musculoskeletal mechanics
Cranial nerves, 10, 273–279. *See also* Muscle innervation pathways
  motor distribution, 277
  sensory distribution, 276
Cremaster, **149**, **152**
Cross-body crunches, 104, 139
Cross-over shin stretch, 247

Deep fascia, 22. *See also* Skeletal muscles
Deep tendon reflex, 28. *See also* Skeletal muscles
Deltoideus, **159**, **171**
Depressor. *See also* Face muscles; Mouth muscles; Scalp muscles
  anguli oris, **53**, **78**
  labii inferioris, **53**, **77**
  septi nasi, **53**, **70**
Diaphragm, **109**, **148**
Digastricus, **30**, **89**, **94**. *See also* Neck muscles; Skeletal muscles
Doorframe chest stretch, 160

Dorsal interossei, **208**, **246**, **270**. *See also* Foot muscles
Double kneeling shin stretch, 247, 261
Dumbbell
  curls, 177, 184
  kickbacks, 177
  overhead extension, 193
  shoulder press, 170
  walking lunge, 155

Ear muscles, 59. *See also* Epicranius; Eyelid muscles; Mastication muscles; Mouth muscles; Nose muscles
  auricularis anterior, **60**
  auricularis posterior, **62**
  auricularis superior, **61**
Elbow joint, 295. *See also* Body movement muscles
Ellipsoid joint, 39. *See also* Synovial joints
Endomysium, 21, 22. *See also* Skeletal muscles
Epicranius, **53**. *See also* Ear muscles; Eyelid muscles; Mastication muscles; Mouth muscles; Nose muscles
  frontalis, **57**
  occipitalis, **56**
  temporoparietalis, **58**
Epimysium, 21, 22. *See also* Skeletal muscles
Erector spinae, **109**
Extensor. *See also* Foot muscles; Leg muscles
  carpi radialis brevis, **196**
  carpi radialis longus, **195**
  carpi ulnaris, **199**
  digiti minimi, **198**
  digitorum, **197**
  digitorum brevis, **246**, **272**
  digitorum longus, **245**, **249**
  hallucis longus, **245**, **251**
  indicis, **204**
  pollicis brevis, **202**
  pollicis longus, **203**
External intercostal muscles, 109
Eyelid muscles, 63. *See also* Ear muscles; Epicranius; Mastication muscles; Mouth muscles; Nose muscles
  corrugator supercilii, **66**
  levator palpebrae superioris, **65**
  orbicularis oculi, **64**

Face muscles, 53. *See also* Epicranius; Eyelid muscles; Mastication muscles; Mouth muscles; Nose muscles
Facial nerve, 54, 273, 274, 278
Fasciculi, 22. *See also* Skeletal muscles
Femoral nerve, 288. *See also* Muscle innervation pathways
Fibrous pericardium, 109
Fibularis. *See also* Leg muscles
  brevis, **245**, **253**
  longus, **245**, **252**
  tertius, **245**, **250**

# THE CONCISE BOOK OF TRIGGER POINTS, 3E
## A PROFESSIONAL AND SELF-HELP MANUAL

Since publication *The Concise Book of Trigger Points* has been translated into over twenty languages and become a best seller worldwide. The content of this new edition has been completely updated and revamped, sharing current research, evidence, and advanced techniques for manual therapy practitioners, as well as simple self-help protocols that the layperson can do at home.

Containing full-color illustrations, it is a compact reference guide, and explains how to treat chronic pain through trigger points – tender, painful nodules that form in muscle fibers and connective tissues. So much pain can be relieved quickly and efficiently with simple trigger point therapy.

The book is designed in an easy reference format to offer useful information about the trigger points relating to the main skeletal muscles, which are central to massage, bodywork, and physical therapy. The first four chapters provide a sound background to the physiology of trigger points, and the general methods of treatment. The following six chapters are organized by muscle group, with the information about each muscle presented in a uniform style throughout. Each two-page spread gives detailed anatomical information, referred pain patterns, plus key trigger point information, practitioner protocols, and self-help information and drawings.

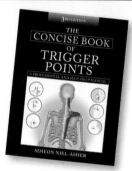

**Simeon Niel-Asher**, B. Phil., B.Sc., (Ost.), who qualified as an osteopath in 1992, uses trigger point therapy in his every day work. He is the inventor of the Niel-Asher technique(TM) for treating frozen shoulder syndrome and was named by the *Evening Standard* newspaper as one of the Top 10 osteopaths in London. He is involved in treating, research, writing, and teaching throughout Europe, the Middle East and the USA.

*"Probably the clearest, most concise and comprehensive presentation of the trigger point picture I have ever read. Eminently practical and clinically useful."*
- Dr. Richard Bachrach, D.O., F.A.O.A.S.M, a US-based doctor who has been at the forefront of trigger points use for more than 20 years.

lotus
publishing

978-1-905367-51-1 (UK) / 978-1-583948-49-1 (US)  |  £18.99 (US) / $29.95 (US)
256 pages  |  paperback  |  350 color illustrations